THE TIMES

STARGAZING

BEGINNERS GUIDE TO ASTRONOMY

Radmila Topalovic and Tom Kerss

ROYAL
OBSERVATORY
GREENWICH

Published by Times Books
An imprint of HarperCollins Publishers
Westerhill Road
Bishopbriggs
Glasgow G64 2QT
www.harpercollins.co.uk

In association with
Royal Museums Greenwich, the group name for the National Maritime Museum,
Royal Observatory Greenwich, Queen's House and *Cutty Sark* 2016
www.rmg.co.uk

This edition for WH Smith 2017

© HarperCollins Publishers 2016
Text © National Maritime Museum
Cover photograph: Milky Way – © NAOJ
Family – Nelson Marques/Shutterstock
Star charts by Wil Tirion – www.wil-tirion.com
Images and illustrations see acknowledgements page 222

The Times® is a registered trademark of HarperCollins Publishers Ltd

The contents of this publication are believed correct at the time of printing.
Nevertheless the publisher can accept no responsibility for errors or omissions,
changes in the detail given or for any expense or loss thereby caused.

HarperCollins does not warrant that any website mentioned in this title will be provided uninterrupted,
that any website will be error free, that defects will be corrected, or that the website or the server that
makes it available are free of viruses or bugs. For full terms and conditions please refer to the site
terms provided on the website.

A catalogue record for this book is available from the British Library

ISBN 978-0-00-796597-7

10 9 8 7 6 5 4 3 2 1

Printed in China by RR Donnelley APS. Co. Ltd

If you would like to comment on any aspect of this book, please contact us at the above address or online.
e-mail: collinsmaps@harpercollins.co.uk

 facebook.com/CollinsAstronomy
 @CollinsAstro

MIX
Paper from
responsible sources
FSC **FSC™ C007454**
www.fsc.org

Find out more about HarperCollins and the environment at
www.harpercollins.co.uk/green

Contents

Planning your Stargazing Session

Start with your Eyes

Taking Pictures

Using Binoculars or a Telescope

Things to See

Constellations and Seasonal Objects

Start Stargazing!

Seasonal Charts

Further Resources and Reading

INTRODUCTION

Many thousands of years ago, our ancestors began to notice the stars. They told stories, traced patterns and eventually came to recognize the annual rhythms of these primeval constellations. This appreciation of the night sky had a profound impact on civilization, propelling humanity into an age of more advanced agriculture, exploration and philosophy. Then, in the seventeenth century, new views of the heavens provided by the telescope helped start a scientific quest for further knowledge, and inspired generations of great painters, musicians and literary artists.

Before light pollution became a problem, everyone had access to astonishingly dark skies, but today many of us live in towns or cities, above which the faint wonders of the Universe are largely obscured. We believe that stargazing is a fantastic hobby well worth pursuing in spite of this obstacle, and in this guide we've outlined ways to get started under urban skies, before moving out to an ideal site in the countryside. We present a selection of choice objects to explore across the sky, with charts to help you find them, and occasional scientific insight. Also covered are many exciting events you can witness with careful planning, and a beginner's guide to the mechanics of our cosmic viewing platform – Earth.

Every astronomical discovery is the result of an unquenchable human thirst to understand our place in the cosmos. Hopefully, the same curiosity has stirred your inner stargazer, and whatever your ambition – whether you wish to tour the planets, experience deep time in the search for distant galaxies, or simply admire the serenity of the night sky as our ancestors did – we hope this book serves you well in taking your first steps into the world of amateur astronomy. Clear skies!

Radmila Topalovic and Tom Kerss
Astronomers at the Royal Observatory Greenwich

THE NIGHT SKY

Earth in Space

ORBIT

We live on a spinning ball of rock – Earth. Formed 4.5 billion years ago, at the same time as the Sun and everything else in our solar system, our planet is the third from the Sun. Mercury and Venus are closer and too hot for liquid water to exist, and therefore too hot for life to exist. Earth, however, has an average temperature of 15°C. Its temperature is partly determined by its orbit: Earth lies in a region around the Sun called the 'habitable zone' where its distance is just right for liquid water to exist. However, its orbital distance is not the only factor that determines whether Earth

can support life. Earth has an atmosphere that strongly affects its overall surface temperature.

Earth takes 365.25 days to orbit the Sun. We have 365 days in the calendar year, and we account for the extra quarter day by adding a day to February every four years. While Earth's orbit is elliptical, it is very close to being circular. An elliptical orbit means that Earth's distance from the Sun varies over the course of a year.

Earth moves around the Sun at an astonishing speed of 30 km per second or 66,600 mph. This is Earth's average speed – because of its elliptical orbit, the orbital velocity

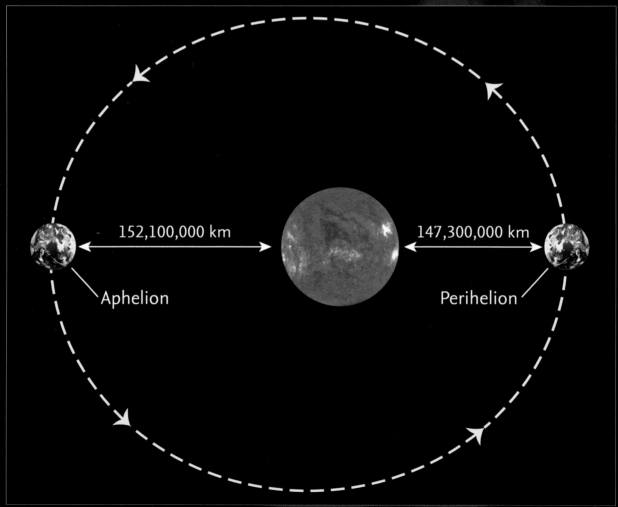

152,100,000 km

Aphelion

147,300,000 km

Perihelion

Earth in orbit around the Sun.

changes slightly as it makes its way around. Earth moves faster at perihelion (our closest point to the Sun) compared to when it is at aphelion (our farthest point from the Sun).

ECLIPTIC

The flat plane of our orbit (imagine a giant disc with a radius of 150 million km or 93 million miles) is called the ecliptic, and it is approximately aligned with the orbital planes of the other seven planets – our solar system is close to flat if we ignore the outer regions beyond Neptune. When we look at the planets moving slowly across the sky night after night, they move roughly in a line through a band of constellations called the Zodiac band. This is our view of the ecliptic extended into space – we are seeing the plane of our solar system.

If there were intelligent beings on another world in our Milky Way galaxy and they were looking at our solar system through a telescope, they would see Earth periodically cross the face of the Sun if they were observing in line with the ecliptic. An alien world would have to be ideally placed to see transits of any planet in the Solar System.

TRANSITS

The inferior planets (those with orbits between Earth and the Sun), Mercury and Venus, can appear to us to cross the face of the Sun on rare occasions – an event known as a transit. The transit method is a technique for finding planets orbiting other stars. As they pass in front of their host star they block some of the starlight. You can observe this using a telescope.

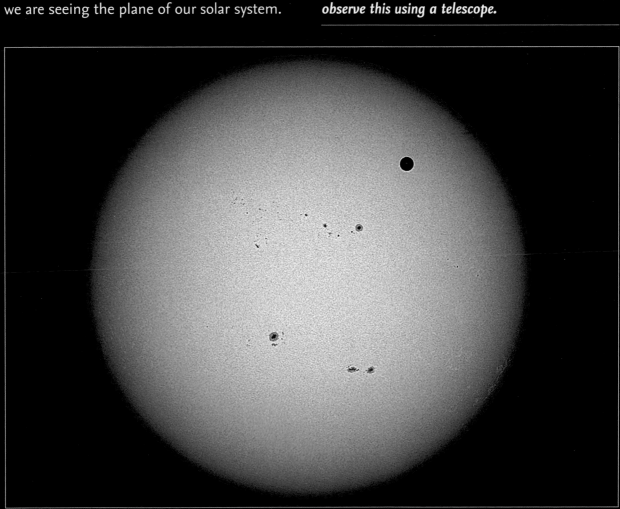

Transit of Venus from Hawaii, 2012.

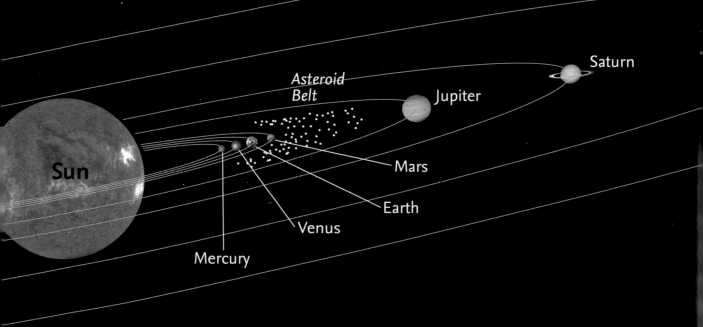

Saturn

Asteroid
Belt

Jupiter

Sun

Mars

Earth

Venus

Mercury

The order of the planets in the Solar System.

Uranus

Neptune

AXIAL TILT

Earth is not upright as it orbits the Sun. Its north–south axis is not perpendicular to the ecliptic. Instead, its axis is tilted by 23.4°, which means our equatorial plane, extended into space, does not align with the plane of the ecliptic.

The tilt affects the weather here on Earth. Countries north and south of the equator experience seasons over the course of the year. In the UK, winter is experienced in December – this is when the northern hemisphere is tilted away from the Sun. At the same time, the southern hemisphere is tilted towards the Sun and so people there enjoy their summertime. Six months later, at the opposite end of our orbit, the northern hemisphere is tilted towards the Sun and people experience summer, whereas those in the southern hemisphere are tilted away from the Sun and are in winter. Near the equator there isn't much seasonal variation, with the temperature staying high throughout the year.

The altitude (its height above your local horizon) of the Sun at midday also varies throughout the year. The Sun reaches its highest point in the sky at midday. In the winter, when we are tilted away from the Sun, the Sun does not reach the same altitude at midday as it did in the summer. We receive less intense sunlight and a shorter period of daylight in winter. Sunrise and sunset times vary across the year because of Earth's axial tilt: in London, for example, the period of daylight swings from 8 hours in mid-winter to 16 hours in mid-summer.

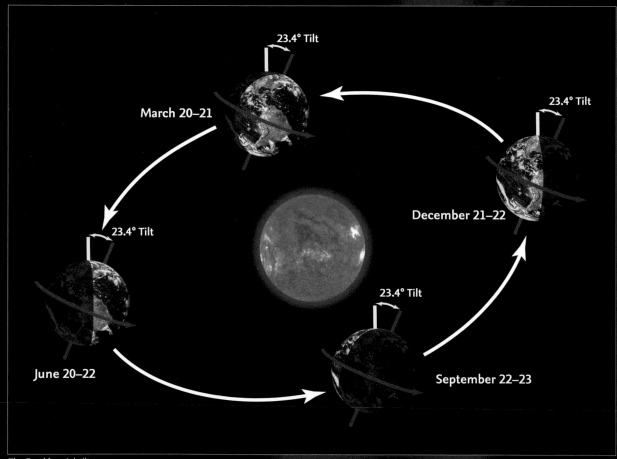

The Earth's axial tilt.

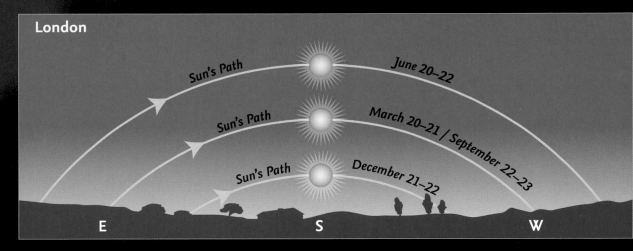

London

June 20–22
Sun's Path
March 20–21 / September 22–23
Sun's Path
December 21–22
Sun's Path

E   S   W

Sydney

December 21–22
Sun's Path
March 20–21 / September 22–23
Sun's Path
June 20–22
Sun's Path

W   N   E

Changing altitude of the Sun throughout the year (top, northern hemisphere and bottom, southern hemisphere).

SPIN (SIDEREAL DAY VS SOLAR DAY)

Earth spins on its axis, taking 23 hours, 56 minutes and 4 seconds to complete one rotation. This is the true rotational period of Earth and we call this the 'sidereal day'. We can see the effects of Earth's rotation when we look at the Sun, the Moon and every other celestial object. The stars continuously move around the north or south celestial pole – in the northern hemisphere, this is marked by Polaris (the North Star). The stars in the night sky take one sidereal day to complete their path around the celestial pole (or to rise, set and then rise again).

However, we follow a 24-hour day called the 'solar day'. We see the Sun rise in the east,

reach its highest point in the sky at midday and then set in the west. This is because Earth spins anti-clockwise (from west to east) when looking down on the North Pole.

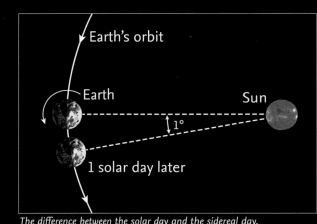

Earth's orbit

Earth

Sun

1°

1 solar day later

The difference between the solar day and the sidereal day.

Star trails – long exposure photographs reveal the Earth's rotation tracing partial circles around the celestial poles.

so the sky appears to move daily from east to west. The solar day is about four minutes longer than the sidereal day. It takes longer because we have to take into account the motion of Earth around the Sun. There are 360° in a circle, 365 days in a year, so Earth covers around 1° of its orbit every day, and it is continuously moving relative to the Sun. From our perspective, the Sun slowly shifts position by 1° from west to east, lagging behind in the opposite direction to its daily cycle. To make up for this 1° motion we wait an extra four minutes for the Sun to reach its highest point in the sky again at local midday.

POSITION OF EQUINOXES

Seasons happen because of the axial tilt of Earth, and mid-winter and mid-summer are marked by the solstices around 21 June and 21 December – these occur when Earth is fully tilted towards or away from the Sun. In between these extremes are two equinoxes around 20 March and 23 September, when Earth is tilted neither towards nor away from the Sun. At the equinoxes, both the northern and southern hemispheres experience approximately 12 hours of daylight and 12 hours of night, but the times of sunset and sunrise still vary depending on your location. The equinoxes mark the start of spring and autumn.

The equinoxes are the points where the equatorial plane of Earth intersects the ecliptic. The axial tilt means the equatorial plane of Earth does not lie parallel to the ecliptic, but they are separated by an angle of 23.4°.

The Sun crossing the celestial equator during the March equinox in London.

LATITUDE AND LONGITUDE

Your local view of the sky depends on your latitude. Lines of equal latitude circle the Earth perpendicular to its north–south axis. Your latitude is how far north or south you are from the equator, in degrees. The latitude on the equator is 0°, Greenwich in London is 51.5° north of the equator, Sydney is 33.9° south, the North Pole is +90° or 90°N and the South Pole is -90° or 90°S. The star Polaris happens to be almost precisely above the North Pole, and its altitude is equal to your latitude in the northern hemisphere. In London, Polaris is 51.5° above the horizon, and at the North Pole it is 90° above.

The Arctic Circle marks a latitude of 66.5°. North of this, the Sun does not set during and around the June (summer) solstice and it never rises above the horizon around the December (winter) solstice. The same

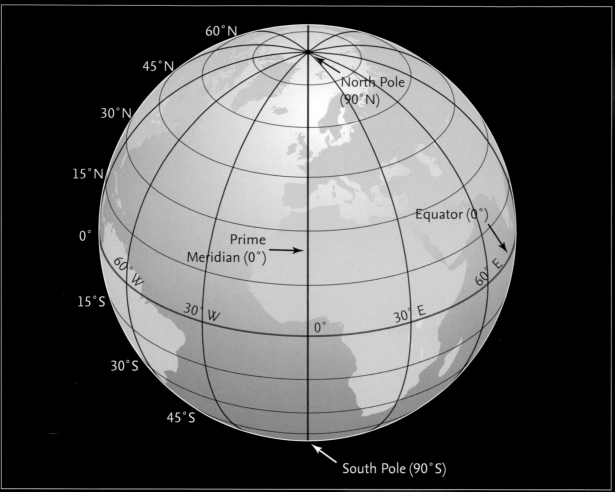

Lines of longitude and latitude.

happens at latitudes south of -66.5° in the Antarctic Circle; however, there the summer and winter periods are reversed.

The lines of longitude are perpendicular to lines of latitude. There are 360 lines of longitude, each 1° apart, all the way around Earth and converging at the North and South Poles. The 0° longitude line – the prime meridian – runs through Greenwich in London, UK, and is marked at the Royal Observatory.

Longitude tells you how far east or west you are relative to Greenwich: for example, Copenhagen is 12.6°E, New York is 74°W. As Earth spins west to east, countries east of Greenwich are further ahead in time, and countries west of Greenwich are behind in time. Local time depends on your longitude: if two towns are 15° of longitude apart, the local time difference is one hour.

Greenwich Mean Time (GMT) – and the Greenwich prime meridian – was internationally recognized in 1884 and adopted across Great Britain some years earlier. Countries across the world set their clocks to their local time zone, which can be whole or half hours ahead or behind GMT (or in a few cases a quarter of an hour ahead) and cover a wide region, ignoring local time differences specific to longitude. Copenhagen is in a time zone one hour ahead of Greenwich, while New York is five hours behind GMT.

Celestial Sphere

Compared with the distance between any two points on Earth, or even the distances to the Moon or other planets, the stars are incredibly remote. Even the nearest star to the Sun – Proxima Centauri – is 9,000 times further from us than Neptune. For simplicity, imagine every object outside the Solar System as being fixed to the inside of an enormous sphere with Earth at the centre. This model is called the Celestial Sphere.

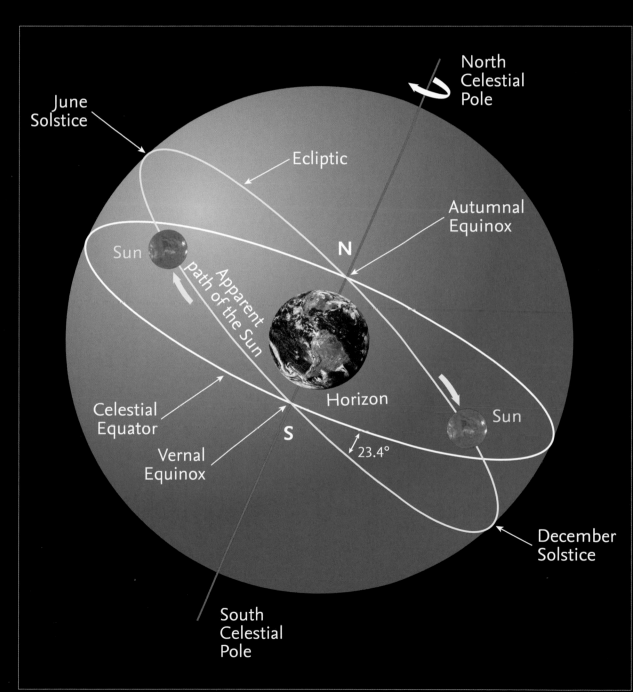

The Celestial Sphere around the Earth.

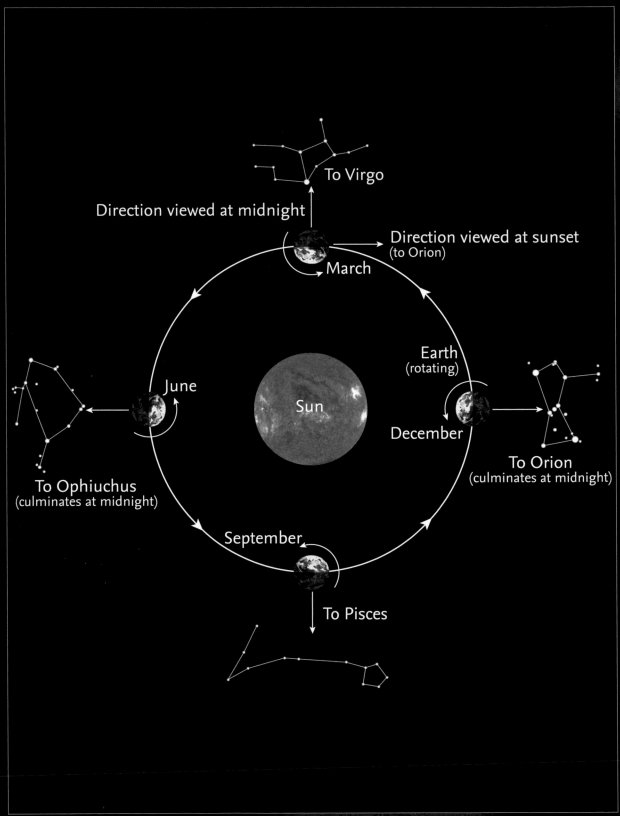

The changing view of the Celestial Sphere as Earth orbits the Sun.
Different constellations oppose the Sun at different times of the year.

THE CELESTIAL POLES

As Earth rotates on its axis, the Celestial Sphere appears to drift in the opposite direction. Earth's axis can be imagined as a line connecting its poles, as well as the poles of the Celestial Sphere. If you stood on the equator, you could theoretically see both celestial poles at the same time, but the vast majority of us see just one pole or the other. For stargazers north of the equator, the celestial pole is very close to the star Polaris (Alpha Ursae Minoris), known as the Pole Star or North Star. It marks the tip of the Little Bear's tail (Ursa Minor).

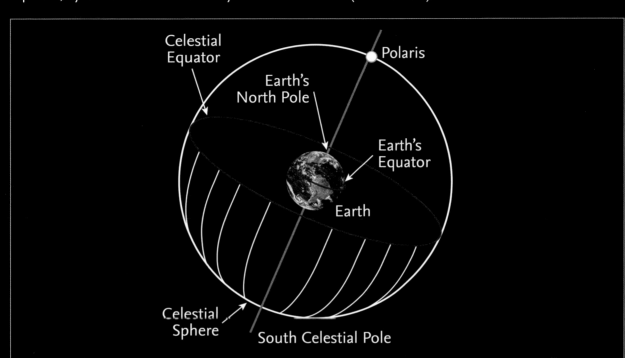

The celestial poles as seen from both hemispheres.

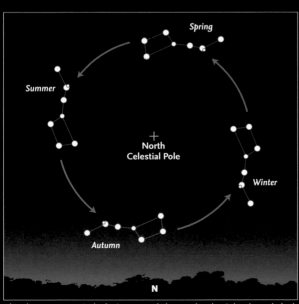

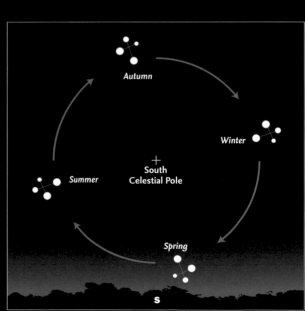

The sky rotates anti-clockwise around the north celestial pole and clockwise around the south celestial pole.
Positions at different times of the year can be found in the seasonal sky charts.

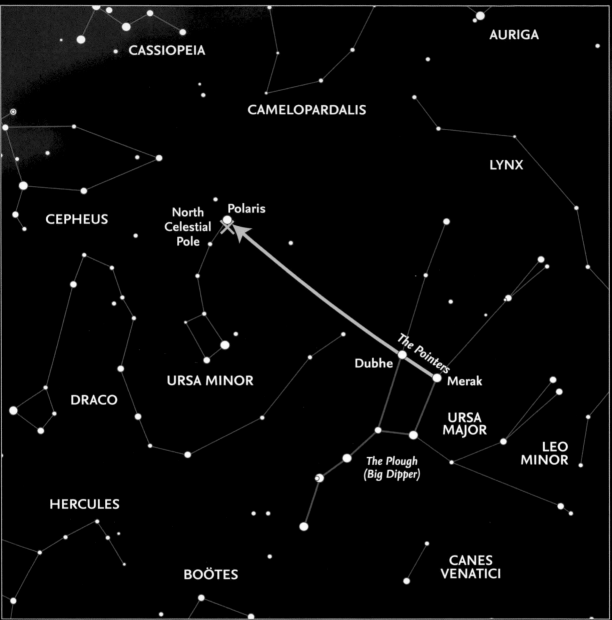

Finding the celestial North pole.

Finding Polaris is easier if you first find the Plough (also called the Big Dipper or Saucepan), made up of the Great Bear's (Ursa Major) three brightest stars. The stars Dubhe (Alpha Ursae Majoris) and Merak (Beta Ursae Majoris) are called the Pointers, since they point the way to Polaris at all times while the Plough circles the pole. By tracing a line between the pair, and continuing this line in your mind's eye across the sky, you will be reliably guided to Polaris!

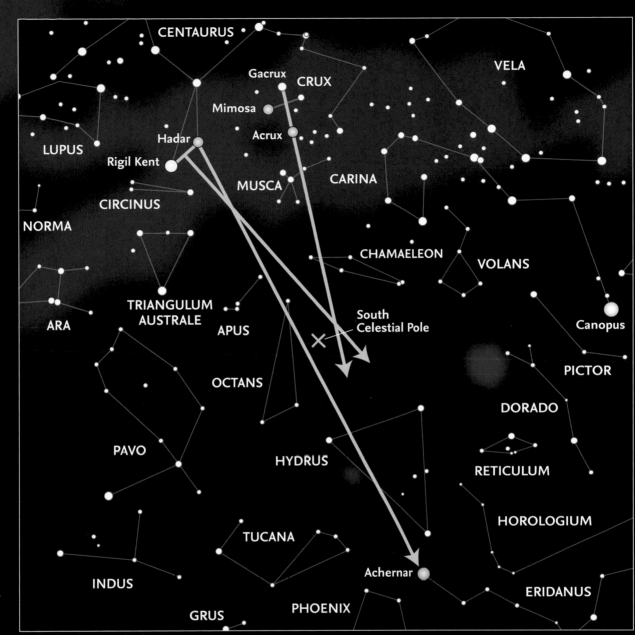

Finding the celestial South pole.

In the southern hemisphere, Sigma Octantis is the current South Star, but it is close to the limit of what you can see without binoculars or a telescope. There is a popular method for finding the south celestial pole using bright pointers. Rigel Kentaurus (Alpha Centauri) and Hadar (Beta Centauri) are the brightest stars in the constellation of Centaurus, and are known as the Southern Pointers since they point the way to the Cross (Crux). Acrux (Alpha Crucis) and Gacrux (Gamma Crucis) mark the Cross's bottom and top. By taking an imaginary line through this pair towards the bottom of the Cross, and another line perpendicular to the one between Centaurus's two brightest stars, you can find the south celestial pole by stopping where those two lines intersect.

CELESTIAL COORDINATES: RIGHT ASCENSION/DECLINATION

We can define any position on the surface of Earth using just two coordinates: longitude and latitude. We can also use two coordinates on the Celestial Sphere to mark the precise position of any star or other object. The equivalent of longitude is called Right Ascension (RA), and the equivalent of latitude is called Declination (Dec). Like Earth, the Celestial Sphere has an equator, dividing the northern and southern hemispheres.

Declination is also measured north or south of the equator in units of degrees. Locations north of the equator are normally given in positive values, so the north celestial pole is at +90°, and locations south of the equator are given in negative values, so the south celestial pole is at -90°. For greater precision, degrees are subdivided into 60 minutes of arc (arcminutes or arcmin) which are further divided into 60 seconds of arc (arcseconds or arcsec), so from pole to pole there are 648,000 arcseconds.

Right Ascension is measured according to Earth's rotation. One solar day is 24 hours long, so RA coordinates are given in hours, minutes and seconds. Like Earth, the Celestial Sphere has a prime meridian of its own, where Right Ascension is zero. This is known as the First Point of Aries, and is defined as the location of the Sun at the vernal equinox (the first day of spring in the northern hemisphere, and the first day of autumn in the southern hemisphere).

When this convention became popular, the Sun was among the stars of Aries, but today the First Point of Aries actually lies among the stars of Pisces, due to precession – the slow wobble of Earth's axis. Because the First Point of Aries changes location, and because every star is gradually changing position, coordinates on the Celestial Sphere always refer to a particular epoch (period of time). The current standard epoch is J2000.0, which means that coordinates are correct as of noon GMT on 1 January 2000. The J2000.0 position of the brightest star in the night sky, Sirius, can be written as: RA: 6h 45m 8.92s, Dec: -16° 42' 58".

ALTITUDE AND AZIMUTH

Positions on the Celestial Sphere are defined by RA and Dec. There are two more coordinates, called altitude and azimuth (Alt/Az), that describe the positions of objects in the sky as they appear on an imaginary dome above your head. The top of the dome, called the zenith, is defined as having an altitude of 90°, and is immediately above you. The horizon is 0°. Azimuth is a bearing measured from north (0°) through east (90°), south (180°) and west (270°). Your local meridian is a line above your head running from horizon at 0° azimuth, through the zenith, all the way to 180° azimuth. The translation from RA/Dec to Alt/Az depends on your location and local time. When looking south from the northern hemisphere, or north from the southern hemisphere, any object reaches its highest altitude for any given night as it crosses the meridian, an event called culmination. If the sky is fully dark, this is the best time to see it.

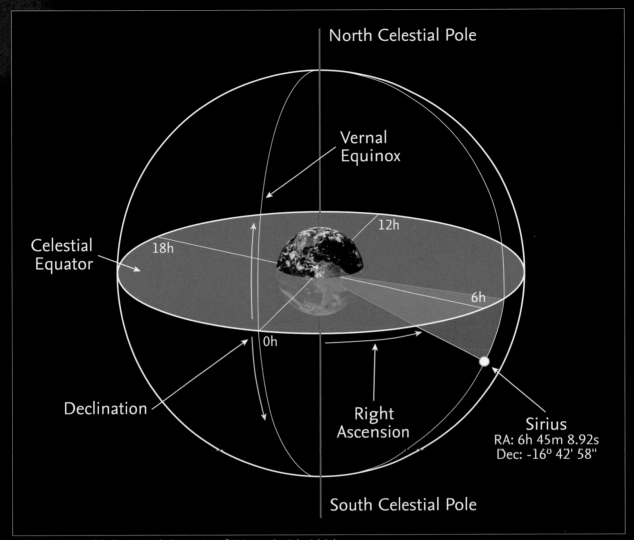

North Celestial Pole

Vernal
Equinox

12h

Celestial
Equator

18h

6h

0h

Declination

Right
Ascension

Sirius
RA: 6h 45m 8.92s
Dec: -16° 42' 58"

South Celestial Pole

Right ascension and declination with the position of Sirius on the Celestial Sphere.

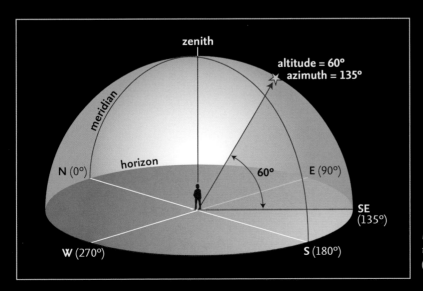

zenith

altitude = 60°
azimuth = 135°

meridian

horizon

N (0°)

E (90°)

60°

SE
(135°)

W (270°)

S (180°)

*Local meridian (line bisecting
the sky east–west) and zenith
(the point directly overhead).*

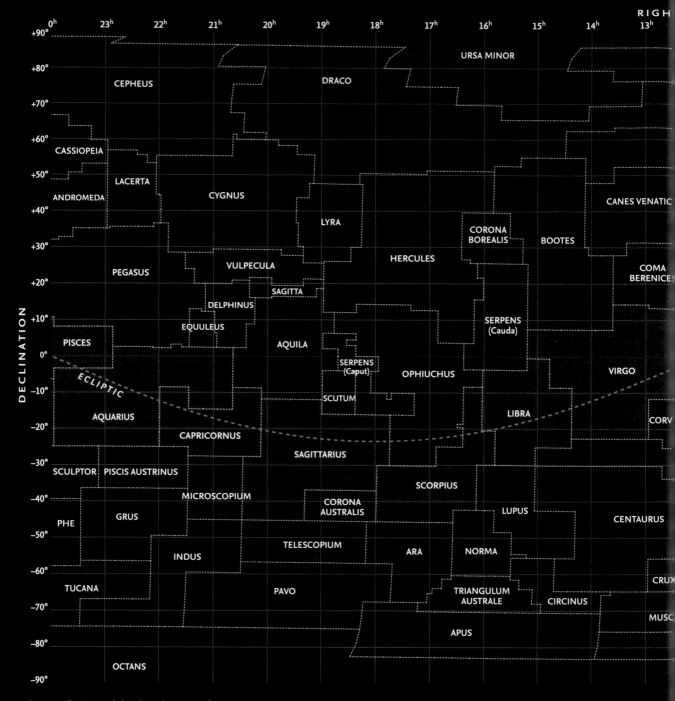

The constellations and their boundaries in a flat projection – the ecliptic is shown running through the zodiacal constellations.

CONSTELLATIONS: BOUNDARIES AND FAMILIES

The ancient Greek astronomer Ptolemy compiled a list of 48 constellations two millennia ago. In total, 88 constellations were in common use by the year 1930, when the International Astronomical Union (IAU) created formal boundaries. Many constellations belong to families, grouped together by theme or mythological story. This diagram illustrates the Celestial Sphere projected onto a flat map, with the equatorial and polar regions

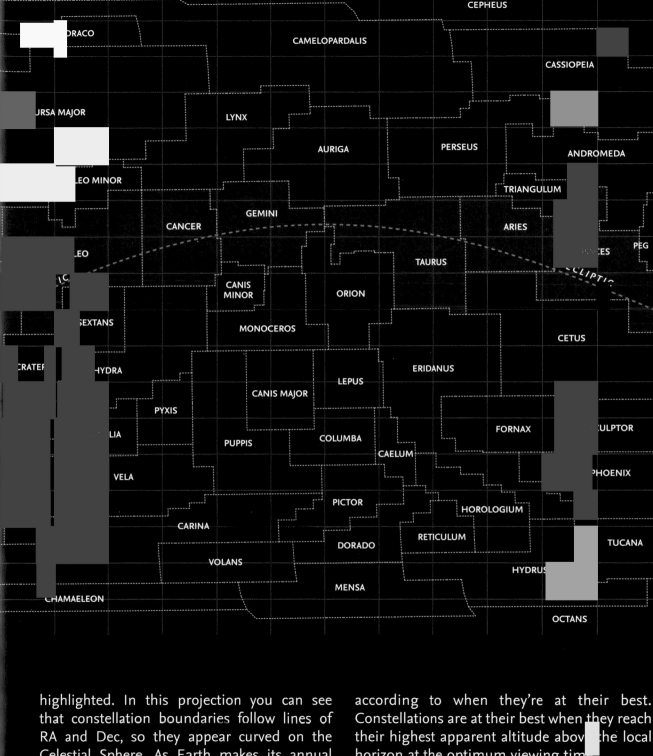

highlighted. In this projection you can see that constellation boundaries follow lines of RA and Dec, so they appear curved on the Celestial Sphere. As Earth makes its annual journey around the Sun, the constellations we see change, and can be grouped by season

according to when they're at their best. Constellations are at their best when they reach their highest apparent altitude above the local horizon at the optimum viewing time.

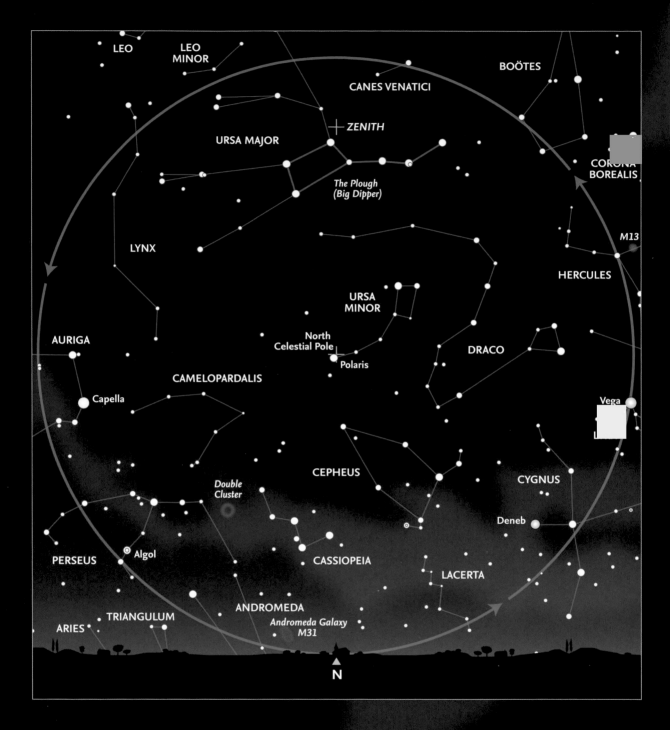

LEO

LEO
MINOR

BOÖTES

CANES VENATICI

— ZENITH

URSA MAJOR

CORONA
BOREALIS

The Plough
(Big Dipper)

M13

LYNX

HERCULES

URSA
MINOR

North
Celestial Pole

DRACO

AURIGA

Polaris

CAMELOPARDALIS

Capella

Vega

CEPHEUS

CYGNUS

Double
Cluster

Deneb

PERSEUS

Algol

CASSIOPEIA

LACERTA

TRIANGULUM

ANDROMEDA

ARIES

Andromeda Galaxy
M31

▲
N

CIRCUMPOLAR CONSTELLATIONS

For each location on Earth, there are a group of circumpolar constellations – those that don't ever set below the horizon on any night of the year. These constellations encircle the celestial pole, but how many remain above the horizon will depend on your latitude. If you are closer to the pole, you will see more circumpolar constellations and fewer equatorial constellations.

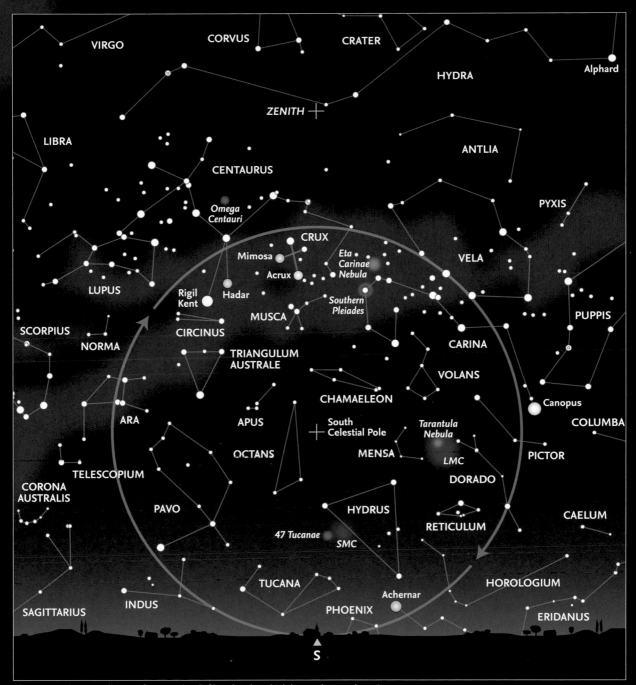

The circumpolar constellations from London (left) and Sydney (right) near the March equinox. Constellations within the circle never set from these latitudes.

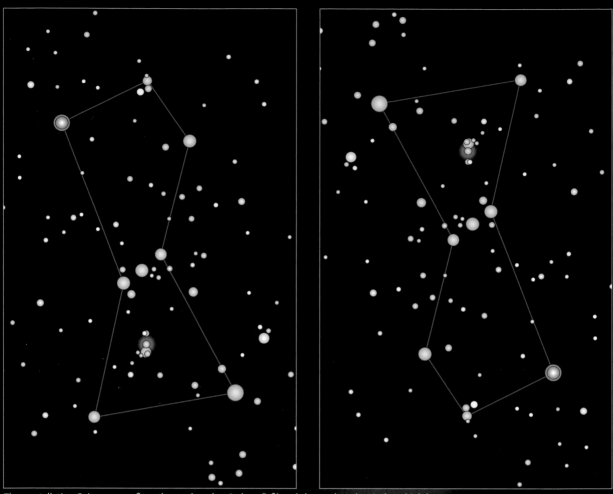

The constellation Orion as seen from the northern hemisphere (left) and the southern hemisphere (right).

EQUATORIAL CONSTELLATIONS

Many of the most famous and recognizable constellations belong to the equatorial region. These constellations are visible throughout most of the world, depending on the time of year. Observers in the northern hemisphere look south towards this region of the sky, while observers in the southern hemisphere look north, so two friends in different hemispheres will see these constellations the opposite way up from one another.

THE ZODIAC

Perhaps the most famous collection of constellations is the Zodiac. These patterns, sometimes referred to as star signs, are of special significance to astronomers. As we make our year-long journey around the Sun, we observe a change in the Sun's position along the ecliptic. This path cuts through all of the twelve zodiacal constellations, as well as a thirteenth – Ophiuchus – according to the boundaries set by the IAU in 1930. It should be noted that constellations and star signs are only loosely related, as signs divide the ecliptic into twelve equal portions of 30°, whereas constellations vary in size according to the positions of the stars. Starting in January, the order of constellations through which the Sun passes is: Sagittarius, Capricornus, Aquarius, Pisces, Aries, Taurus, Gemini, Cancer, Leo, Virgo, Libra, Scorpius, and Ophiuchus (which is not officially a zodiacal constellation).

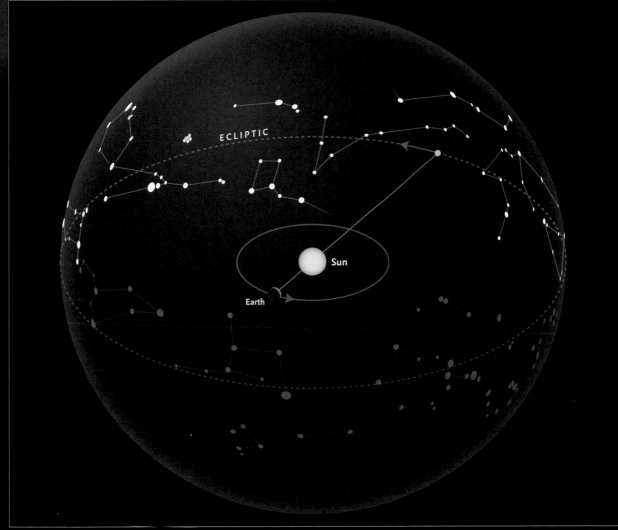

The zodiacal band tracing the ecliptic.

The Solar System

THE SUN

The Sun is a yellow dwarf star, its brightness is too low for it to be a giant star. In the core of the Sun, hydrogen atoms are fused to form helium at a rate of about one hundred trillion trillion trillion reactions per second! The energy released by this process is light of all different types. Dwarf stars do not end their lives in a supernova explosion like giant stars. As the Sun is our nearest star, it is one of the very few astronomical objects that can be seen during the day. On average, Earth and the Sun are separated

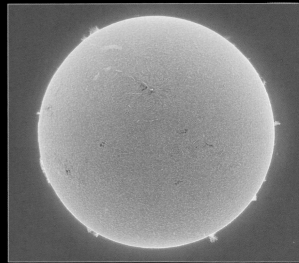

Our active star, a ball of hydrogen and helium gas.

by just over 149 million km (just under 93 million miles), but its influence is seen and felt; the Sun's intensity is so great it floods our nitrogen-rich atmosphere with light. Blue light is preferentially scattered in all directions, obscuring our view of the stars during the day.

THE SUN IS A G2 STAR

Stars are classified according to their temperature – they each have a spectral type. They are grouped into the following classes: O, B, A, F, G, K, and M, where O-type stars are the hottest and appear blue and M-type stars are the coolest and appear red. The Sun is a G2-type star – a yellow dwarf with a surface temperature of 5,500°C. (See table on page 53).

THE MOON

The Moon is a natural satellite of Earth and the only other world to have been visited by humans. Released from Earth's crust in a catastrophic collision billions of years ago, the Moon would have once loomed much larger in our skies, glowing from the intense heat of great seas of lava all over its surface. Over billions of years, it has cooled and solidified, and moved much further away. The Moon is still moving away from Earth, but at such a slow rate that stargazers will never notice it.

Today, the Moon completes one orbit of Earth every 27.3 days. Because of the motion of Earth and the Moon around the Sun, the length of one lunar orbit is different from the length of time between one New Moon and the next, which is 29.5 days. During this period, known as a lunar synodic month, the familiar phases of the Moon take their turns to appear.

The Moon rotates on its own axis once every 27.3 days, resulting in the misleading illusion that it doesn't rotate at all, since we can only see one side from Earth. Only the Apollo mission astronauts have seen the far side of the Moon with their own eyes.

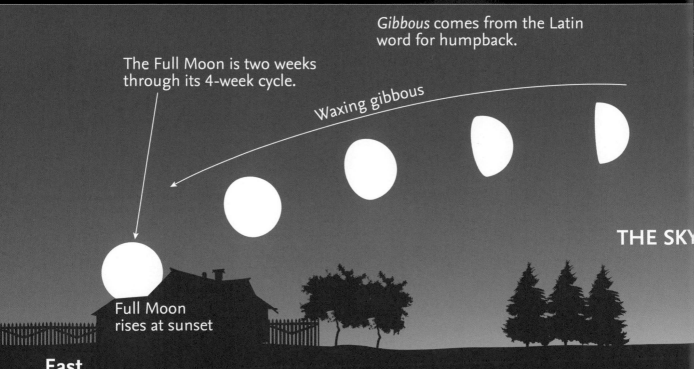

Gibbous comes from the Latin word for humpback.

The Full Moon is two weeks through its 4-week cycle.

Waxing gibbous

THE SKY

Full Moon rises at sunset

East

First half of the lunar cycle showing phase and position of the Moon at sunset.

Waning Crescent Moon and Venus at sunset from the European Southern Observatory at Paranal, Chile.

The First Quarter Moon is one week through its 4-week cycle.

The first two weeks of the cycle of the Moon is shown below by its position at sunset on 14 successive evenings. As the Moon grows fatter from new to full it is said to wax.

Waxing crescent

AT SUNSET

New Moon is invisible near the sun

South

West

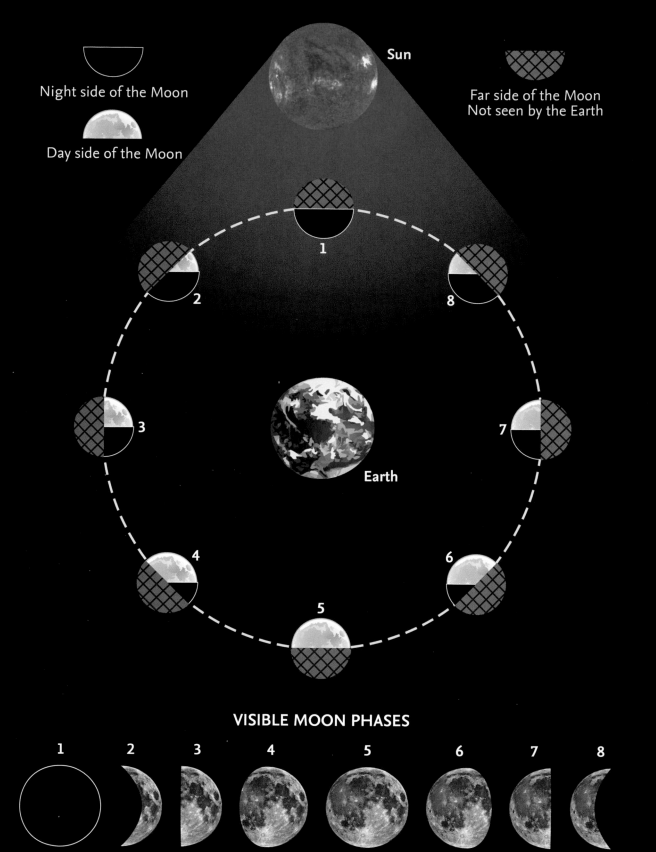

Night side of the Moon

Day side of the Moon

Sun

Far side of the Moon
Not seen by the Earth

Earth

VISIBLE MOON PHASES

1	2	3	4	5	6	7	8
New	Waxing Crescent	First Quarter	Waxing Gibbous	Full	Waning Gibbous	Last Quarter	Waning Crescent

The rest of us have to rely on images from robotic orbiters, which have mapped the entire lunar surface in detail.

The distance between the Moon and Earth also changes, this is because the Moon's orbit is not perfectly circular: it varies between 356,500 km (221,500 miles) at lunar perigee and 406,700 km (252,700 miles) at lunar apogee. When the Full Moon occurs around lunar perigee, it appears slightly larger and somewhat brighter than average – this is an event known as a supermoon.

PLANETS
Eight known planets orbit the Sun, separated by many millions of miles. In ancient Greece, astronomers dubbed them *asteres planetai* – the wandering stars. (See pages 36–37).

Because of the way the Solar System formed, the planets' orbits are closely aligned to the plane of the Sun's equator, so they never stray far from the ecliptic in the sky. Since it is closer to the Sun, Mercury can be particularly difficult to observe after sunset or before sunrise unless your horizon is clear of obstructions like trees and buildings. Venus, which appears considerably brighter than Mercury and further from the Sun, is more forgiving. Mars and the four gas giant planets can at times appear opposite the Sun, visible all night long.

MINOR PLANETS
A minor planet is a natural object orbiting the Sun that is not a planet or a comet, such as an asteroid or dwarf planet. Because of their smaller size, minor planets often look like stars, even in large telescopes. Hundreds of thousands of minor planets have been discovered and catalogued, though only a few, such as Pluto, Ceres and Vesta, are accessible to the amateur astronomer.

ASTEROIDS
Asteroids are leftover pieces from the formation of our solar system: fragments of irregularly shaped rock and metal largely concentrated in a region between the orbits of Mars and Jupiter. Though they may not appear impressive to the observer, they're a sobering reminder of the real dangers presented by space rocks, such as the one responsible for the demise of the dinosaurs 66 million years ago. The largest easily observable asteroid, Vesta, is around 60 times wider!

COMETS
Comets are small objects made of dust and ice, and are temporary visitors from the distant Kuiper Belt – a doughnut composed of dwarf planets and comets, and the home of Pluto – or the Oort Cloud, a huge bubble of icy rocks thought to surround the Solar System.

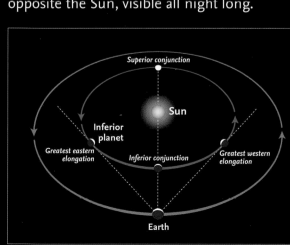

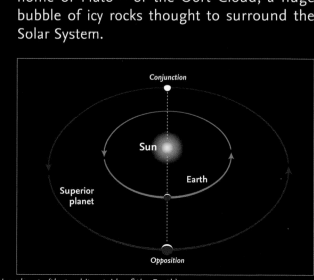

The inferior planets (that orbit between the Sun and the Earth) and superior planets (that orbit outside of the Earth).

Sun

Mercury

Venus Earth

Mars

Ceres
Vesta

*Asteroid
Belt*

Jupiter

Comet

Our solar system (not to scale).

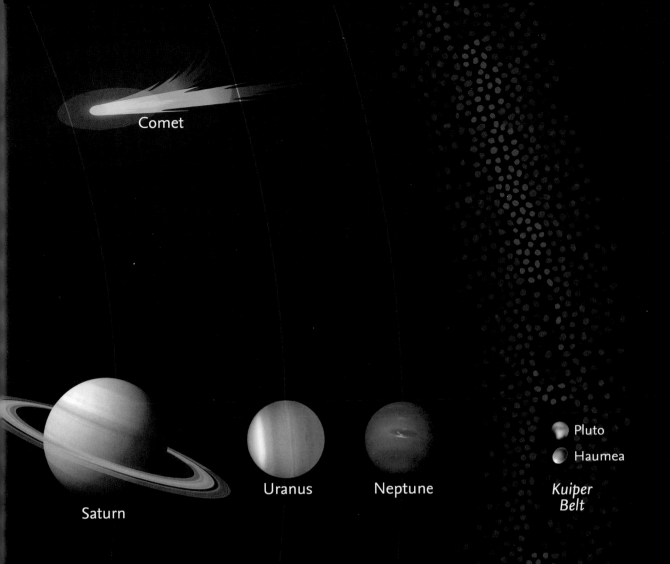

Comet

Saturn

Uranus

Neptune

Pluto

Haumea

Kuiper Belt

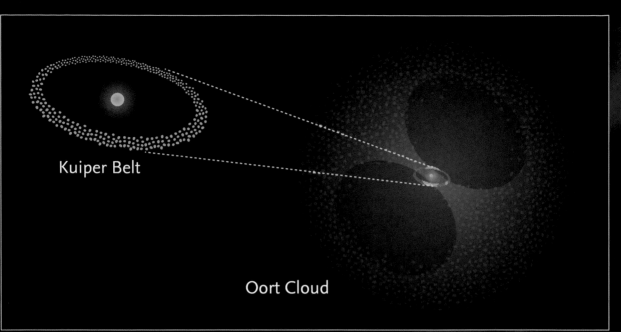

The Kuiper Belt and the Oort Cloud.

Many comets make a single approach to the Sun, growing a spectacular tail of gas, dust and ice crystals millions of kilometres in length before soaring away never to be seen again. Others have short, stable orbits, such as Halley's Comet, which orbits the Sun every 76 years, or Encke's Comet, which takes just 3.3 years.

Comets that have made multiple trips around the Sun have usually exhausted their volatile surface material and no longer produce tails, but several of these familiar visitors have contributed clouds of debris that give us dazzling annual meteor showers.

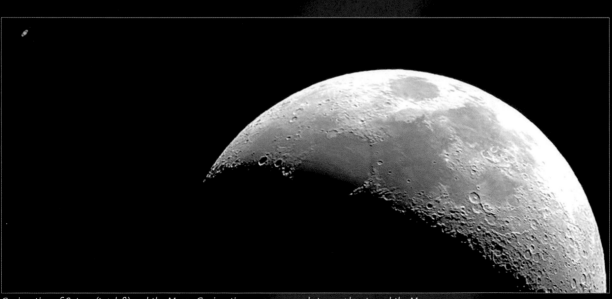

Conjunction of Saturn (top left) and the Moon. Conjunctions are common between planets and the Moon (when they appear close together in the sky).

Transient Events

METEORS AND METEOR SHOWERS

Meteors, sometimes called shooting stars, are rocky or icy fragments burning up many kilometres above us in Earth's upper atmosphere. Most meteors appear for just a fraction of a second, so spotting these fleeting events is often down to luck.

Meteors typically disintegrate about 80 km (50 miles) above us, travelling over 16 km (10 miles) per second! Occasionally, a larger fragment will enter Earth's atmosphere and produce an exceptionally bright meteor. Meteors brighter than any of the planets are called fireballs. Bright fireballs are thrilling to see, and they can show faint or pronounced colour, as well as leaving behind a lingering train.

Every night there are hundreds of opportunities to see individual meteors, but showers occur on an annual basis and can be predicted according to Earth's orbit. Examples include the Geminids, which peak in the second week of December, and the Perseids, seen every year in August. Showers deliver a much greater rate of meteors than normal, because they're the result of comets leaving enormous clouds of icy debris throughout the inner Solar System. For example, Halley's Comet is responsible for the annual Eta Aquariids in May and the Orionids in October.

ECLIPSES

Eclipses of the Sun and Moon are among the most remarkable shows put on by nature. Feared by our more superstitious ancestors, we now understand they are rare alignments, casting shadows on a colossal scale. We don't see eclipses every month because the Moon's orbit is inclined by about 5° from the ecliptic. There are two points, called nodes, where the orbit of the Moon and the ecliptic cross each other. If the Moon is near these nodes when

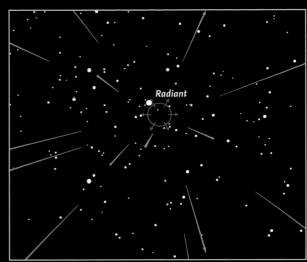

The radiant of the Geminids meteor shower (from the constellation Gemini).

its phase is new or full, eclipses will be visible somewhere on Earth.

Lunar eclipses can be seen when the Moon (in its Full Moon phase) passes through Earth's shadow during its orbit. The penumbra, or outer region, casts a pale grey shadow on the Moon, and the umbra at the centre is considerably darker with a red hue – the Moon is reflecting the light of every sunset and sunrise on Earth.

Solar eclipses happen at New Moon when the Moon's shadow falls on Earth. When we are under this shadow, all or part of the Sun's disc (its visible surface) is covered up, resulting in a period of relative darkness during the day. Since looking directly at the Sun is dangerous, it's best to use projection methods to see solar eclipses indirectly. Solar eclipses are also more difficult to see, since the shadow of the Moon is relatively small, and most of Earth's surface is covered with water.

Eclipses can be total or partial. During a total lunar eclipse, the Moon is immersed entirely within the umbral shadow of Earth at eclipse maximum. During a total solar eclipse, the Sun's disc is completely covered

Successive stages of a lunar eclipse – the red hue is due to atmospheric refraction and filtering of sunlight reaching the Moon.

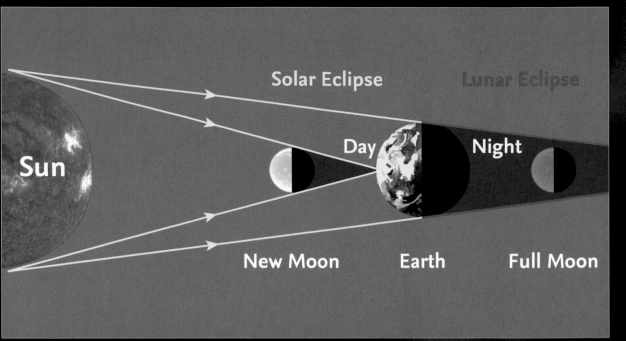

Solar Eclipse   Lunar Eclipse

Sun   Day   Night

New Moon   Earth   Full Moon

Positions of the Sun, Earth and Moon during a solar and a lunar eclipse.

by the Moon, and the extensive solar corona can be seen for a few minutes.

Partial eclipses are more common. In the case of a partial lunar eclipse, the Moon only partially enters Earth's umbral shadow. During partial solar eclipses, the Sun's disc looks like a crescent or truncated circle.

Due to the changing distance from Earth to the Moon, an annular solar eclipse can happen when a total solar eclipse occurs around lunar apogee. In these cases, the Moon is too far away to totally cover the Sun, so a ring of light can be seen.

AURORAE

Aurorae, or polar lights, are a spectacular natural phenomenon. They occur when our magnetic field collects energetic particles from the solar wind, accelerating them into the atmosphere near the magnetic poles, where they excite and ionise atoms and molecules of oxygen and nitrogen, causing them to glow.

Perspective

ANGULAR SIZE

The angular size of an object is a measure of its width or diameter in degrees. For example, a building looks bigger to you the closer it is, and so its angular size is greater. If you move further away from the building, its angular size decreases.

The angular sizes of celestial objects are smaller than a degree, so astronomers use smaller units called 'arcminutes' and 'arcseconds'. The angular resolution limit of the human eye is 1 arcminute: this means we can differentiate between two objects 0.3 metres apart, at a distance of 1 km. Our eyes alone can't resolve (distinguish) objects with an angular separation smaller than 1 arcminute.

The Moon's orbit around Earth is elliptical, which means its distance changes as it makes it way around. This affects its angular size: when it is furthest away from us, the lunar disc covers 29.3 arcminutes

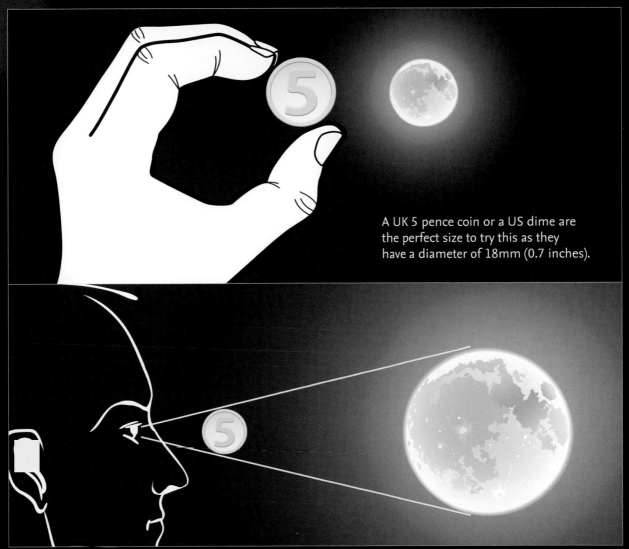

A UK 5 pence coin or a US dime are the perfect size to try this as they have a diameter of 18mm (0.7 inches).

Covering the Moon with a small coin. The angular size of the Moon stays constant as it rises and sets across the sky.

(0.5°) in the sky, and when it is closest to us, it covers 34 arcminutes and it becomes a supermoon. The Moon and the Sun have almost identical angular sizes, which is why we occasionally see a total solar eclipse.

MOON ILLUSION

You may have noticed that when the Moon is low in the sky it looks really big. When the Moon is high it doesn't look as big – this strange effect is an illusion created by your brain. In fact, the Moon has the same angular size at all altitudes over the course of the night or day.

When the Moon is low in the sky, your brain compares it to the surrounding landscape. Relative to the buildings and trees around us, the Moon looks bigger than usual. When you see the Moon high in the sky, the landscape is out of your field of view, so it looks smaller. You can break through the illusion by holding out a small coin at arm's length and use it to cover the Moon at both low and high altitudes – you'll see there's no difference.

SIZE AND DISTANCE

The closest spiral galaxy to us is called Andromeda and it is 2.5 million light-years away, equivalent to 24 million trillion km. A light-year (ly) is the distance light covers in a year. A text message to Andromeda travelling at the speed of light – 300 million metres per second or 670 million miles per hour – would take 2.5 million years to reach the galaxy. It has an angular size of 3°, making it over six times as wide as the Full Moon. We can't see the full extent of Andromeda with the naked eye or small telescope as it is too faint, but on a dark clear night you should be able to make out the brighter central region.

OPTICAL ILLUSIONS

Another optical illusion is caused by an effect on light called refraction. Light slows down as it enters our atmosphere and this causes it to change direction. You can observe this effect yourself by placing a pencil or straw inside a glass of water. You will see that it looks broken – this is because your brain assumes light always travels in straight lines and it cannot account for the fact that light is being refracted by the water.

When you look at the Moon it appears to be a little higher in the sky than it actually is. This is due to the refraction of light from the Moon as it travels through the atmosphere. The effect of refraction means that stars are not quite where they appear to be, and that the Sun can still be seen just above the horizon even after it has technically set.

Refraction of light has a beautiful effect on the Full Moon during a lunar eclipse – it turns a blood-red colour. When sunlight travels through Earth's atmosphere and bends towards the Moon, blue light is filtered out but red light continues through to the Moon.

Refraction of light through a glass of water.

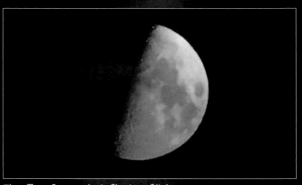

The effect of atmospheric filtering of light on a low altitude Moon.

LIGHT AS A WAVE

Light travels as a wave from the Sun to Earth. The Sun emits different types of light or radiation: radio waves, microwaves, infrared, visible light, ultraviolet, X-rays and gamma rays. The hottest stars in our galaxy are more intense in the high-energy short wavelength part of the spectrum, whereas cooler stars emit more low-energy long wavelength light.

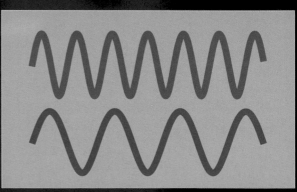

Blue and red light waves. Blue light has a shorter wavelength than red light.

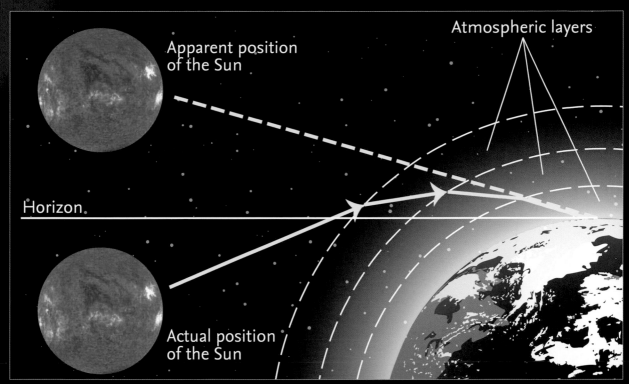

Apparent position of the Sun

Atmospheric layers

Horizon

Actual position of the Sun

Apparent and actual positions of the Sun due to atmospheric refraction of light.

THE SUN FROM MARS

On Mars, the sky is pink: this is because its atmosphere is full of dust particles that are slightly larger than the particles in Earth's atmosphere. They scatter longer wavelength red light and allow blue light through unimpeded. Sunrise and sunset on Mars appear blue because of this effect.

SCATTERING OF LIGHT

Have you ever wondered why our sky is blue? An effect called Rayleigh scattering is responsible. If sunlight passes through a glass prism, we see the colours of the spectrum (a rainbow) – this is called 'dispersion'. Red light has a longer wavelength than blue light, so red light from the Sun passes through our atmosphere

Sunset on Mars – image taken by the NASA Opportunity rover.

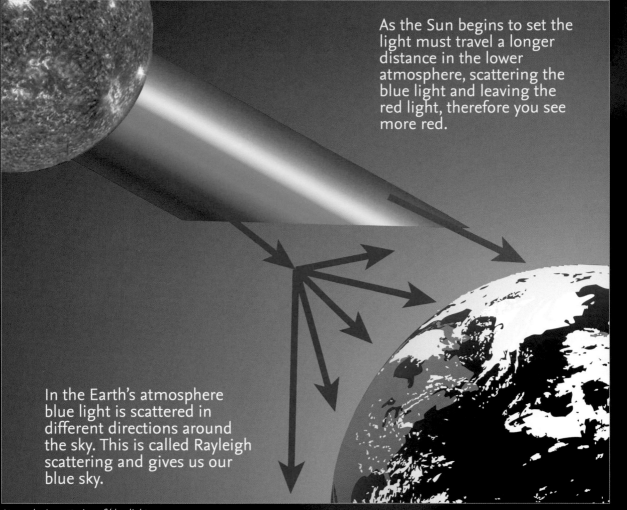

As the Sun begins to set the light must travel a longer distance in the lower atmosphere, scattering the blue light and leaving the red light, therefore you see more red.

In the Earth's atmosphere blue light is scattered in different directions around the sky. This is called Rayleigh scattering and gives us our blue sky.

Atmospheric scattering of blue light.

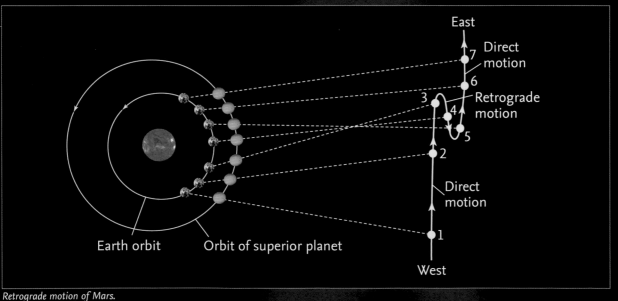

Retrograde motion of Mars.

Mars undergoing retrograde motion – Mars increases in brightness due to its closest approach to Earth.

without any problems but blue light is scattered by the nitrogen and oxygen molecules in our atmosphere. This happens because the size of the gas molecules corresponds to the wavelength of blue light. Blue light is scattered in all directions, turning our sky a vivid blue colour.

When the Sun is low on the horizon, its light takes a longer journey through the atmosphere compared to when it is highest in the sky and almost all of the blue light is scattered out, leaving behind predominantly red-orange light that eventually reaches our eyes. When we look back towards the rising or setting Sun, we see it glow red.

RETROGRADE MOTION

The planets move from west to east through the twelve constellations that make up the Zodiac. Planets closer to us, like Venus and Mars, appear to move faster through the sky than more distant planets, like the gas giants. Earth is also moving around the Sun faster than the planets orbiting beyond us: Mars, Jupiter, Saturn, Uranus and Neptune. This means that

over time, we approach these planets and then overtake them in our faster orbit – when this happens, it seems as if the planets stop in their tracks and appear to move backwards. Astronomers call this U-turn effect 'retrograde motion'.

You can find out when retrograde motion is going to happen from astronomical calendars. Then you could chart or photograph the position of that planet in the sky over time, relative to the stars close to it.

MORNING AND EVENING STARS

Mercury and Venus (the inferior planets) also exhibit retrograde motion, but this is harder to see because it happens either side of the Sun. Venus loops in the sky every 584 days when it overtakes Earth. Before this, it can be seen to the left (east) of the Sun and is visible after sunset when it is known as the evening star. After overtaking Earth, Venus can be seen on the right (west) of the Sun and it becomes the morning star, visible just before sunrise.

PLANNING YOUR STARGAZING SESSION

Star trails over Mount Elbrus in the Caucasus Mountains, Southern Russia

THE HUMAN EYE

The eye is an extraordinary instrument, adapted and refined by millions of years of evolution to produce vivid, detailed images of the world around us. Tuned to the peak output of the Sun – a range of colours we call visible light – our eyes, in tandem with our brains, interpret a useful slice of the electromagnetic spectrum, while compensating for movement and changing light levels, to provide us with an effortless sensation of sight. Gazing at the Galaxy wasn't an evolutionary priority for the eye, so we have developed tools to see further into the night sky.

Your eyes have two vision systems, based on different kinds of photoreceptor (light sensitive) cells. Photopic vision makes use of cells called cones. These are largely concentrated very near the centre of the retina, where your vision is sharpest.

Photopic vision is used in well-lit conditions, but when light levels are very low our scotopic vision system gets to work. Scotopic vision makes use of photoreceptor cells called rods. These are concentrated towards the edges of the retina, where you experience peripheral vision, and make up over 90 per cent of the nearly 100 million photoreceptors in each of your eyes. Rod cells come in just one variety, so they can't differentiate between colours on their own. Their location on the retina also means they have lower resolution and less c... these drawbacks are outweighed by the huge increase in sensitivity that they offer. When rod cells are active, the brain can produce meaningful images in drastically reduced light.

BECOMING 'DARK ADAPTED'

Each eye collects and focuses light through a system of lenses known as the pupil. Maybe you've stumbled into your bathroom in the middle of the night and turned on the light to be greeted by a reflection of yourself with supernaturally wide, dark pupils. Whether they're open in darkness, or closed, your eyes automatically adapt to try and compensate for the lack of light by dilating the pupil, effectively increasing its aperture. This results in greater light grasp, allowing more light to illuminate the cells on the retina. When a bright light source dazzles you, your pupil shrinks again. The maximum size that the pupil can dilate to declines with age, but it always improves your stargazing experience, and it is the first step to becoming dark adapted.

THE CAMERA COMPARED TO THE EYE

The camera and the eye are very similar in the way they interact with light. Both have an aperture – in the eye it is the pupil – to allow light through. The size of the aperture can change depending on light levels. Light then passes through a lens and the rays converge to a point (called the focal point) producing an inverted image. The image is then detected and interpreted, in the camera by a CCD chip and a computer, and in the eye by the retina and the brain.

STAYING ADAPTED

Becoming fully dark adapted takes at least 30 minutes, but readjusting to bright light takes only a few seconds. Even a relatively soft light, such as a phone screen on its dimmest setting, can set you back for several minutes.

Here are two tried-and-tested methods to keep your eyes set to stargazing mode:

1. Avoid bright lights! The best way to stay dark adapted is to avert your eyes from all artificial lights.

2. Buy a red torch. Rod cells are insensitive to lower energy red light, so even bright red lights can be used without spoiling things.

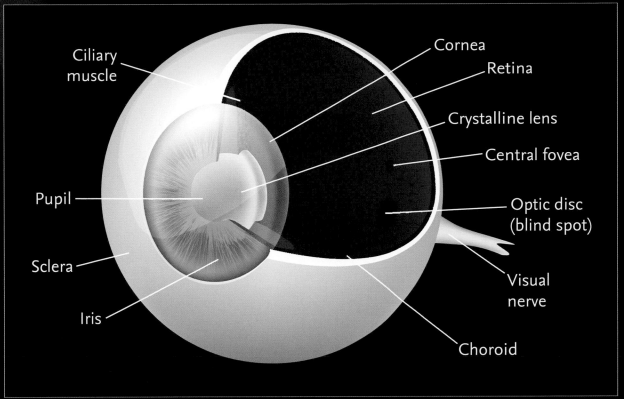

The anatomy of the human eye.

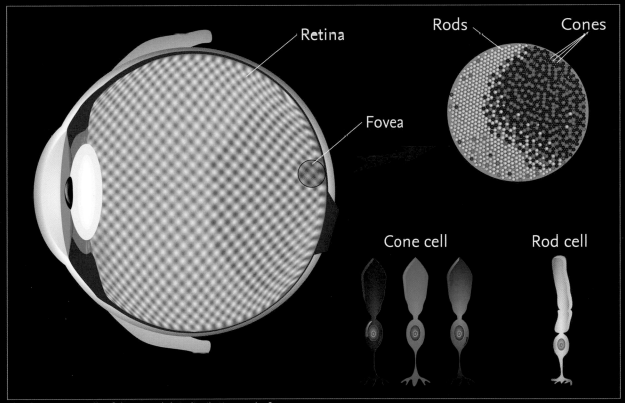

The photoreceptor cells of the eye and their distribution on the fovea.

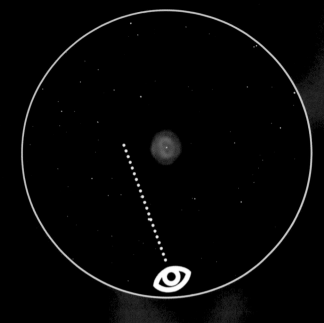

The effect of averted vision.

Some well-designed astronomy apps for tablets and smartphones even have a 'red light mode' so they can be used in the field.

AVERTED VISION

Cone cells are so concentrated near the centre of the retina that rods are nearly absent, resulting in a low-light blind spot at the centre of each eye. Averted vision is a technique used to overcome the blind spot by using the rest of the retina, where rod cells are found. Put simply, this involves looking slightly to the side of what it is you want to see. Try it indoors, with the lights off, by looking at a sheet of white paper across the room. When your eyes are dark adapted, you'll find that the paper appears brighter if you focus your gaze to one side of it. Averted vision is very effective when trying to spot extremely faint objects in the night sky.

SEEING COLOUR

It may seem like enjoying colour in the night sky is a lost cause. After all, colour sensitivity

Mars as it appears to the unaided eye (left) and through the Great Equatorial Telescope at the Royal Observatory Greenwich (right).

Class	Conventional description	Actual apparent colour	Temperature (Kelvins)
O (Meissa)	blue	blue	≥33,000 K
B (Spica)	blue to blue-white	blue-white	10,000 – 30,000 K
A (Sirius)	white	white to blue-white	7,500 – 10,000 K
F (Procyon)	yellowish-white	white	6,000 – 7,500 K
G (Sun)	yellow	yellowish-white	5,200 – 6,000 K
K (Arcturus)	orange	yellow-orange	3,700 – 5,200 K
M (Antares)	red	orange-red	≤3,700 K

The colours of stars. The Sun's surface temperature is around 5,800 K or 5,500 °C.

is severely reduced when the eye is fully dark-adapted. But the reality is that, although it is extremely subtle, colour in the night sky can be seen.

Every star has a real colour dictated by its surface temperature. Red stars are relatively cool, whereas blue stars are hot. From a distance, the Sun would appear as a yellow-white star, but we are so close to it that it seems overwhelmingly bright and therefore appears to be white. One way to enhance your experience of colour is to view colourful double stars. When seen together, the contrast between pairs of stars enhances their apparent colour. You can also slightly defocus your telescope so the stars are blurred and the colours separate further.

In low light conditions, the eye's overall sensitivity to light shifts in favour of the blue end of the spectrum, because rod cells

Colourful double stars in Cygnus – Albireo (left) and 61 Cygni (right).

have a peak sensitivity for blue-green, whereas photopic vision with cone cells peaks in green. This is known as the Purkinje effect. Before colour sensitivity disappears entirely, it's possible to observe faint green and blue hues in certain types of nebulae.

The planets have the most obvious colours in the night sky: there is the unmistakable orange tint of Mars, the complex mixture of colours in Jupiter's stormy cloud tops, and even Neptune's deep blue skies can be seen with a modest telescope.

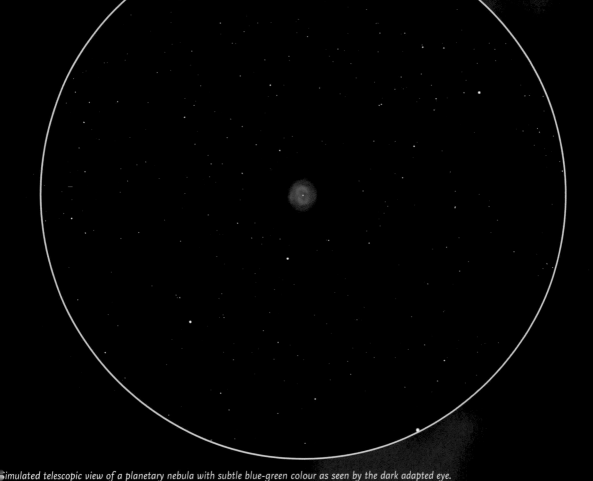

Simulated telescopic view of a planetary nebula with subtle blue-green colour as seen by the dark adapted eye.

Using Starcharts

MAGNITUDES

Magnitude refers to how bright an object appears in the sky. Hipparchus may have been the first to introduce this system, initially dividing the known stars into six groups. The brightest stars were said to be of the first magnitude, and the faintest stars of the sixth. In 1856, astronomer Norman Pogson formalized the system and declared that a sixth-magnitude star should be exactly one hundred times fainter than a first magnitude star. Each increment in magnitude is equal to approximately 2.5 times more brightness.

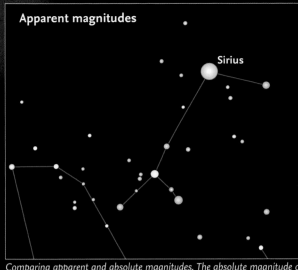

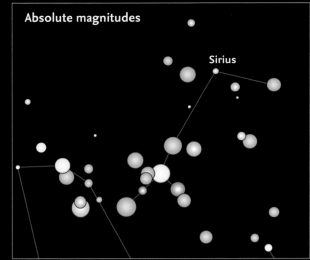

Comparing apparent and absolute magnitudes. The absolute magnitude of an object is its apparent magnitude if it was 32.6 light-years away from Earth.

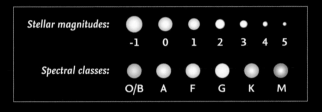

Stellar magnitudes:	-1	0	1	2	3	4	5
Spectral classes:	O/B	A	F	G	K	M	

Key to stellar magnitudes.

Astronomers extended the scale in both directions to include every object in the sky. Not surprisingly, the Sun lies at the extreme negative end of the spectrum, with an apparent magnitude of only -26! The brightest planet seen from Earth is Venus, which can reach a maximum apparent magnitude of -4.89, while remote Neptune reaches 7.78, falling below the limits of the human eye in ideal conditions (about 6.5). About 9,500 stars fall within this limit.

PLANISPHERES

Every astronomer benefits from having a modern-day astrolabe – a planisphere. They are simple to use, relatively cheap and help you to plan your stargazing nights throughout the year without a computer, tablet or phone. You must use one that is correct for your latitude: for example, stargazers in northern Europe, the northern USA and Canada should use one set at latitude 51.5°N (the latitude of Greenwich). The Collins Planisphere is a useful aid for your stargazing sessions, available for the northern hemisphere.

A planisphere is a disc of two layers – the top layer has a transparent window and it rotates above the bottom layer, which is a star chart. It is difficult to project the spherical sky on a flat plane, so there will be some distortion, like there is in a flattened map of the world. Typically a planisphere set at 51.5°N will have the north celestial pole at the centre of the star chart, marked by Polaris. Around Polaris are circles representing declinations 30° apart (see section Right Ascension and Declination).

Also on the star chart is a line (usually dotted) representing the ecliptic. The planets, the Moon and the Sun can be found gradually moving along this line in the night sky relative to the background stars. To see the positions of the planets and Moon on any night you can use Stellarium or another star chart app.

Around the edge of the star chart are the months of the year, split into days. Outside of this are celestial longitudes in degrees and their astronomical equivalents, right ascensions (RA), in hours. The point of 0° (on the celestial sphere) is 0 hours RA, and all stars have right ascensions between 0 and 24 hours. You can also see the relative brightness of stars, represented by their size on the star chart.

Modern-day planisphere published by Collins, set at 50°N.

This book contains a set of 16 seasonal charts to help you identify constellations and the best time of year to see them where you live. You will find the northern hemisphere charts starting on page 182 and the southern hemisphere charts on page 198. Another useful book is Collins annual *Guide to the Night Sky* where suitable for your location.

ASTRONOMICAL SOFTWARE

Stellarium is a free open-source virtual planetarium. It is very easy to use and allows you to play with the night sky from the comfort of your home.

Start with setting your location and the date and time. You can fast-forward and rewind time. Fast-forward through an entire day and you will see the sky rotating anti-clockwise around the north celestial pole if you are in the northern hemisphere, or clockwise around the south celestial pole in the southern hemisphere.

Switching the atmosphere on and off allows you to pinpoint the positions of fainter objects that might be visible in the dawn or during dusk. You can bring up the celestial sphere as a grid, where the lines of right ascension converge at the north or south celestial pole. The azimuthal grid shows you you local lines of altitude and azimuth, with lines of azimuth converging at the zenith (at an altitude of 90° above the local horizon).

Stellarium is a must for every beginner – it is a fun and interactive way to help plan your stargazing sessions. Other astronomical software that we recommend for beginners are Celestia and Starry Night.

Celestia is free and allows you to 'fly' to chosen objects – you can see the planets move relative to each other, and then take a journey to an extrasolar planet (a planet around another star). You can even leave the Milky Way entirely. Celestia will show you spacecraft and satellites currently in our solar system and it has an extensive catalogue of asteroids, comets, stars

and galaxies. Using both Stellarium and Celestia will enable you to have the whole planetarium experience on your computer.

SMARTPHONES AND TABLETS

There are many astronomical apps available for your smartphone or tablet. Most apps now have night-vision mode – very useful in a stargazing session when you don't want to lose your dark-adapted vision. Many also allow you to see what's up in front of you when you point your phone in a particular direction, using its built-in compass and accelerometer.

Pocket Universe for iOS reveals an interactive star map that changes with the direction of your phone. It also shows the lunar phases and deep sky objects, and stores a catalogue of 10,000 stars. An Android alternative is Sky Map, which has a time-travel feature so you can see the sky change over time. SkySafari 4 is bursting with content that is accessible offline. It has a catalogue of 120,000 stars, as well as star clusters, nebulae and galaxies. You can use it to search for asteroids, comets and satellites – these look like bright dots moving slowly across the sky. It also has a time-travel feature so you can see the sky as it was 100 years ago or 100 years into the future. Skyview is another very popular app that uses the camera on your phone or tablet to show which objects are up in the sky in real-time. It will even pick up the ISS and the Hubble Space Telescope passing overhead. Skyview works without WiFi, which makes it really useful if you're out in the countryside or even on an aeroplane. You can find more popular astronomy apps in the Further Resources section.

Clear Skies

WEATHER

The first thing you should check when planning your stargazing session is the weather. See Further Resources for tips on how to find your local forecast.

If you are hoping to see deep sky objects like faint nebulae and galaxies, or if you'd like to see some globular star clusters, you need to choose a clear cloud-free night, preferably with low humidity. Away from city lights, you may be fortunate to see noctilucent clouds once the Sun has set – these are very thin wispy clouds of ice particles that form more than 75 km (46 miles) high in the atmosphere. They become visible when sunlight illuminates them, while the lower layers of the atmosphere are in the shadow of Earth. Noctilucent clouds are best seen throughout the summer months if you live at a latitude between 50° and 70° north or south of the equator.

If it is partially cloudy and you have a camera with you, you might want to take some atmospheric shots of the Moon in between wisps of cloud. If you are using binoculars or a telescope, you will find the front ends gather condensation once you move your equipment somewhere warmer. Do not use your hand or glove to wipe the mist off – always use a lens cloth to clean the eyepiece and aperture, and keep the lens cloth clean. Any accumulated dust or grime on the cloth may scratch the lens or glass.

SUNRISE

The time of sunrise and sunset depends on your latitude and longitude. You can find your local times from weather forecast websites or from the website timeanddate.com. Sunrise and sunset depend on the time of year: in the northern hemisphere, after the December solstice, sunrise takes place progressively earlier, and sunset takes place later and later up until the June solstice, after which the reverse happens and the period of daylight gradually decreases; in the southern hemisphere, the opposite occurs.

On a clear day the sky can turn many shades of red, in sharp contrast to the rest

Noctilucent clouds over Alberta, Canada.

of the blue sky. You can start looking for bright planets like Venus, Mercury, Jupiter and Saturn close to these times and they often lead to a spectacular photograph. You may also catch a Waxing Crescent Moon close to sunset or a Waning Crescent Moon close to sunrise.

There are successive periods of darkness that occur after sunrise and before sunset:

1. 'Civil twilight' is the brightest form of twilight and begins when the centre of the Sun is less than 6° below the horizon.

2. 'Nautical twilight' occurs when the centre of the Sun is between 6° and 12° below the horizon. Many years ago sailors would use the stars visible at this time for navigation.

3. 'Astronomical twilight' happens when the centre of the Sun is between 12° and 18° below the horizon – this is the darkest form of twilight and the time to start looking for faint deep sky objects.

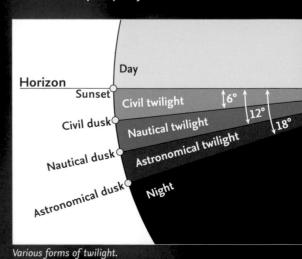

Various forms of twilight.

Sunset over Brighton, UK.

In the UK, full darkness is achieved during the long nights of winter. During the late spring and summer, depending on latitude, the darkest conditions are either nautical or astronomical twilight, from northern to southern latitudes respectively. The city of Rio Gallegos in Santa Cruz, Argentina (latitude 51°S) experiences the same hours of twilight at opposite times of the year to London. See Further Resources for websites where you can calculate your local hours of night, which depend on your latitude.

MOONRISE

It is important to know whether the Moon will be above the horizon during your stargazing session, even if you're not planning on looking at the Moon. When trying to see deep sky objects such as planetary nebulae (dying stars) and galaxies, or a meteor shower, the Moon becomes a source of natural light pollution. Ideally the Moon should be below the horizon when looking at the stars. If it is up, a thin waning or waxing crescent won't interfere with your stargazing as much as a Full Moon.

The Moon rises in the east and sets in the west every day like the Sun, due to Earth's rotation. It also moves slowly west to east relative to the background stars – it covers about 13° of sky every day. This means it rises around 50 minutes later each day.

When planning your stargazing session check when the Moon is up and its phase. A great website to use is timeanddate.com, which will also tell you when the Moon will cross your local meridian.

Moonrise shown as a sequence of photographs.

SEEING

'Seeing' is related to the weather and you will notice that this term comes up a lot in astronomy. Seeing is a measure of atmospheric turbulence: temperature changes in the atmosphere lead to convection currents. Warmer air rises, cools down and then descends. This rapidly moving air distorts light trying to pass through it, and affects what you see through your binoculars or telescope.

Poor seeing means that features like craters on the Moon will seem to shimmer, as if you are looking at them underwater. Major professional ground-based telescopes have adaptive optics to produce a clear, sharp image without any atmospheric distortion. It is a good idea to note the seeing in your observing log (see page 220) every time you head out, particularly if you take photos, so you can compare images of the same object under different seeing conditions. The Antoniadi scale is a commonly used method of measuring seeing:

I. Perfect seeing, without a quiver.

II. Slight quivering of the image with moments of calm lasting several seconds.

III. Moderate seeing with larger air tremors that blur the image.

IV. Poor seeing, constant troublesome undulations of the image.

V. Very bad seeing, hardly stable enough to allow a rough sketch to be made.

You can try to avoid poor seeing by choosing your location carefully. Avoid observing on the pavement because it gives off more heat than grass. Also try not to observe close to hills – air moves upwards over hills, which leads to high levels of turbulence and poor seeing.

The famous rhyme 'Twinkle, Twinkle, Little Star' refers to the apparent twinkling of stars in the night sky – however, stars don't actually twinkle in space. The twinkling happens because of atmospheric distortion. Planets don't appear to twinkle because they are closer to Earth and therefore have a larger angular size and their image isn't affected as much by the moving atmosphere.

TRANSPARENCY
Another factor to consider before stargazing is the transparency of the atmosphere. Dust, smog and other particulates reduce the

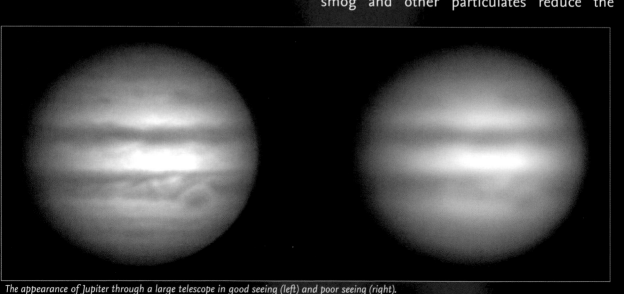

The appearance of Jupiter through a large telescope in good seeing (left) and poor seeing (right).

Sky glow over London.

transparency and make it harder to see faint objects like nebulae and galaxies. When it rains, the transparency increases as the air becomes clearer. If the wind is low, particulates tend to settle in the atmosphere, so although the seeing is good the transparency will be poor.

Professional terrestrial telescopes are located on the tops of mountains and extinct volcanoes in regions where the climate is very dry. Here, the effects of the atmosphere are significantly reduced and astronomers receive the clearest, sharpest images possible from Earth. Ideally telescopes should be placed hundreds of kilometres high in the vacuum of space, like the Hubble Space Telescope. However, a telescope as big as the ten-metre Keck telescope on Mauna Kea in Hawaii would be very difficult to launch and assemble in space.

LIGHT POLLUTION

Weather permitting, you can stargaze whenever and wherever you want, but it gets a lot easier and more interesting if you leave the city and head out into the countryside or somewhere more remote. Light pollution in the city looks like a brown-orange glow that is more intense close to the horizon, and it is the combined effect of scattered light from buildings, cars and streetlights.

However, there is still plenty to see from the city despite light pollution. You will find a list of things to see in the city in this book, though fainter objects will be masked by this artificial glow. The limiting apparent magnitude in city skies is +3: this means that objects fainter than +3 cannot be seen. The limiting magnitude of the eye is +6, so to see these fainter objects you must leave the city and choose a spot as far as possible from any sources of light.

START WITH YOUR EYES

In this chapter you will find tips for observing in urban areas and beyond to the darkest regions in the world, without any equipment – just your eyes.

Urban Skies

You can start tonight if the weather is clear! Hop outside and look at the Moon, and try finding some asterisms (small patterns of stars) or constellations. You can use the observing log provided in this book (see page 220) to record what you see and the local conditions. You can find more information about the objects from Stellarium or another star app.

In the city, the night sky is affected by particulates or smog, which reduces the transparency, and there will also be light pollution to contend with. You should avoid looking at objects low in the sky close to the horizon – the light has to pass through a thicker layer of atmosphere and the effects of sky glow are greater at low altitudes. When planning your next stargazing trip, choose a time when your chosen object or constellation is high in the sky, as close to the meridian as possible. Stellarium will provide you with the hour angle of an object – the closer the hour angle is to either 24 or 0, the closer the object is to the meridian.

Avoid standing next to streetlights – aim for somewhere darker and if possible where there are no tall trees or buildings obscuring your view. In darker regions avoid looking at anything bright to allow your eyes to become dark adapted – this will improve your ability to see fainter stars.

STARS
The limiting magnitude in the city is about 3 to 4, which means you may not see objects

Ideal sky compared to light polluted sky – Orion.

fainter than this. You will see the brighter stars in constellations, and some open clusters will be visible, but galaxies and nebulae may not be visible without binoculars or a telescope. Try finding the Pleiades (M45) in Taurus or Ptolemy's Cluster (M7) in Scorpius. You should aim to look for constellations and fainter objects when the Moon is below the horizon.

MOON AND PLANETS

Try following the monthly phases of the Moon – you could sketch what you see. Look for the dark seas and some of the bright craters such as Copernicus or Tycho. Occasionally the Moon eclipses another object such as a distant star or a planet – this is called an occultation.

There are five planets visible to the naked eye: Mercury, Venus, Mars, Jupiter and Saturn. Mercury can be very difficult to see as it is often very close to the Sun and not as bright as Venus, which stands out in brighter dusk or dawn skies. Planetary conjunctions are a beautiful sight – this is when two or more planets appear to be very close to each other in the sky. You can track the motions of the planets through the Zodiac band: Mercury, Venus and Mars cover daily angular distances of about one to two degrees across the sky, moving a lot faster than the gas giants. You can find information about lunar, solar and planetary events in the Further Resources section.

Waxing Crescent Moon from Greenwich, London.

COMETS AND METEORS

Occasionally, bright comets are visible from the city. Comet Hale-Bopp reached perihelion in April 1997 and was incredibly bright, visible from light-polluted cities with an extensive dust tail covering 40 degrees of the sky. Strong meteor showers can be seen on a clear night from urban areas, but the view is always more spectacular from rural darker regions. Try spotting meteors from the Perseids in August, the Orionids in October and the Geminids in December (famous for its occasional fireballs). More information about significant celestial events can be found in the annual *Guide to the Night Sky* (HarperCollins).

Comet Hale-Bopp with Andromeda below right, seen from Croatia, 1997.

Rural Skies

STARS
Outside of the city, the number of stars visible to the eye changes from hundreds to thousands, so look for the faintest stars in recognisable constellations. Remember to allow at least 20–30 minutes for your eyes to adapt to the dark. Depending on the site, the limiting magnitude will be around 5 to 6. Make a note of the faintest objects you can see in your observing log, and this will help you plan your next stargazing session. You may want to compare the effects of light pollution by counting how many stars you can see in a particular constellation both in the city and the countryside.

You may also see a faint band of light across the sky – this is the Milky Way. There are lots of star clusters to find in darker skies. From the northern hemisphere look for the Beehive Cluster (M44) in Cancer and the Double Cluster (NGC 869 and NGC 884) in Perseus – these are young, open clusters. From the southern hemisphere look for challenging clusters such as M48, a faint open cluster in Hydra, and M22, a globular cluster in Sagittarius. You will need to use averted vision to see these in very dark skies.

SOLAR SYSTEM
The planets really stand out relative to dark rural skies: look for the pink-orange glow of Mars and the yellowish colour of Jupiter and Saturn. Comets and meteors are much easier to see here: meteor showers light up the sky in a spectacular natural firework display. Try looking for the fainter Quadrantids from Boötes. Peak activity lasts for only a few hours, but you could see over a hundred meteors per hour.

NEBULAE AND GALAXIES
Search for fainter objects like diffuse nebulae containing young developing stars and other galaxies. Look towards the sword of Orion to find the Orion Nebula (M42) – a vast star factory. It will look like a faint pink cloud, with two stars either side making up the rest of the sword. In the southern hemisphere lies the magnificent Carina Nebula, a brighter cosmic 'maternity unit' full of stars much hotter and bigger than the Sun.

Try finding the satellite galaxies of the Milky Way – the Large and Small Magellanic

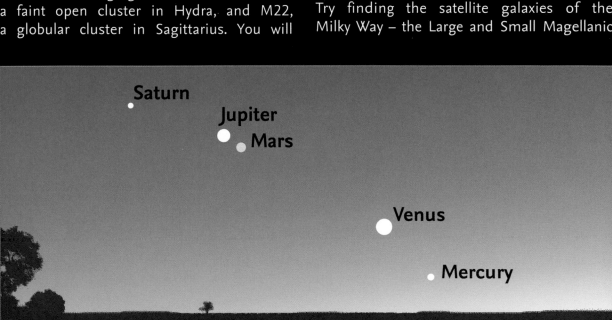

Planets visible to the naked eye.

The Milky Way taken from Black Rock Desert, Nevada, USA.

Clouds (LMC and SMC), only visible in the southern hemisphere. These are diffuse in appearance, the LMC covering about ten degrees of the sky. The LMC contains the Tarantula Nebula, a very intense star-forming region.

The Darkest Skies in the World

Once you are confident finding your way around the night sky, try a more challenging site – head for a designated dark sky reserve or park. You can find a list of sites here: darksky.org/idsp/reserves. In the UK use www.darkskydiscovery.org.uk.

Dark sky sites must meet certain criteria set by the International Dark-Sky Association (IDA) to qualify. Gold tier sites have minimal lighting restricted to the horizon and the visual limiting magnitude is 6.8 when skies are clear. The Milky Way is visible along with faint meteors (that may occur outside of peak meteor shower times) and the very faint zodiacal light – this is sunlight scattered off interstellar dust grains within our solar system and is notoriously difficult to see.

In silver and bronze tier sites, the limiting magnitude drops down to 5.0. In these sites the Milky Way can just be seen, along with the closest spiral galaxy to us, Andromeda. Use averted vision to see the brighter nucleus: the outer parts of the galaxy are too faint to see but if they were visible the galaxy would be six times bigger than the Full Moon. Averted vision will help you to see the fainter objects and they are always easier to see when they are at culmination. Remember to take a red torch with you to navigate your way around in the dark without losing your night vision. You may also want to take binoculars or a telescope with you and a DSLR camera – long-exposure photographs will reveal the colours of planets and emission nebulae.

The night sky over Yosemite National Park, USA.

The night sky over the Atacama Desert, Chile.

TAKING PICTURES

Capturing the Night Sky

The night sky is a unique low-light scene, which makes astrophotography a challenging and addictive form of photography. A tripod is essential since exposure times will vary between a few seconds and a couple of minutes. However, even a tripod will not deliver perfectly sharp stars unless the exposure is minimized, because Earth is constantly rotating. You can only fully compensate for this rotation by using an equatorial tracking platform – a device that turns your camera at precisely the same rate as Earth, but in the opposite direction and aligned with its axis.

Without this set-up, try photographing star trails (stars moving across the sky as the Earth spins) with long exposures, ranging from 10 minutes to composites of several hours. The constellations, Milky Way and transient events, such as aurorae, meteor showers or passes of the ISS, are all excellent subjects for the novice astrophotographer.

We recommend taking a course in astrophotography to develop an in-depth understanding of the theory before splashing out on expensive equipment.

Choosing Cameras, Lenses and Accessories

Astrophotography is much easier with total control over your camera's exposure settings, decent optics and access to high quality digital files. For these reasons, we recommend starting with a Digital Single Lens Reflex (DSLR) camera. There are many major brands offering good DSLRs on the market, or you can find one second-hand at a very reasonable price. We recommend making sure the model you choose has a live view function, as this will become invaluable for focusing.

Top Tips

Whether or not you opt for equatorial tracking, here are several tips to help ensure dramatic images of the night sky:

Use a remote shutter or timer so there's no vibration when the shot is taken.

Manually focus on a bright star (or the Moon) using live-view and 10x zoom. Get the star to be as point-like as possible, or focus on the edge of the Moon (known as the limb).

Your lens won't be able to autofocus on stars, so leave it on manual. If you're out on a cold night, try using a small piece of tape to hold the focus, because thermal expansion inside the lens may push it slightly off the best focus position.

If you want to capture the colour of stars accurately, use daylight white balance. Shoot your images in RAW and as a large JPEG. You can use the JPEG to preview and then process from the RAW. If you're taking long exposures, your camera's sensor will become warmer than the ambient temperature. Give it time to cool down between exposures. Exposure times are a matter of experimentation. A bright Northern Lights display can produce a stunning image in just 5–10 seconds. The Milky Way usually needs longer exposures (30–60 seconds). Meteor showers are sporadic, so why not open your shutter for a few minutes and see what you get?

If your lens is F/4 or faster, you can take well-balanced photos using ISO 400–800.

Attaching a DSLR to a telescope is a simple way to take high-magnification images of the Moon and planets.

A DSLR camera mounted on a tripod.

A live view of the Moon from a camera connected to a small telescope.

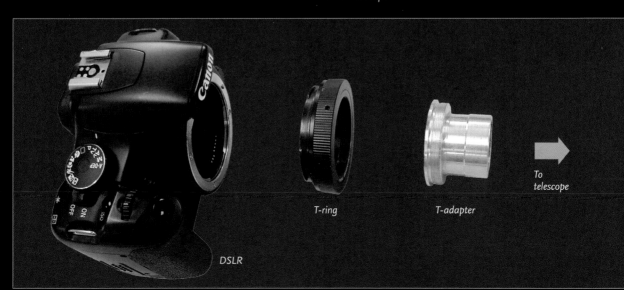

T-ring

T-adapter

To telescope

DSLR

DSLR camera, T-ring and T-adapter to set up a DSLR camera for a telescope.

The Orion Nebula taken with a phone camera (left) compared to the Hubble Space Telescope (right).

The small constellation Delphinus (dolphin) photographed from London (multiple 30 second exposures, ISO 400, 50mm F/1.8 lens, composite)

Lunar eclipse photographed from London (1 second exposure, ISO 400, 50mm F/1.8 lens, composite).

USING BINOCULARS OR A TELESCOPE

Binoculars and telescopes are used mainly to increase the amount of light that reaches the eye to help us detect fainter objects in the night sky. Magnification is not as important as you might think – a lot of the night sky is best seen at magnifications that seem surprisingly low.

Binoculars

Binoculars are a comfortable way to increase the number of stars you can see from a few thousand to several hundred thousand, and bring many deep sky objects into view. They're an ideal introduction to aided stargazing, and we recommend starting out this way before progressing to a telescope.

CHOOSING BINOCULARS
Every pair of binoculars is rated according to its magnification (power) and the diameter of its aperture in millimetres. For example, a pair of 10x50 binoculars has two 50 mm lenses and provides a magnification of 10x. Because of their generous light grasp and compact size, 10x50 binoculars are an excellent portable option.

For slightly deeper views, try 15x70 or even 20x80 binoculars. At this size, many more objects become visible, but these binoculars are heavier and difficult to hold for long periods. Mount your binoculars on a tripod using an

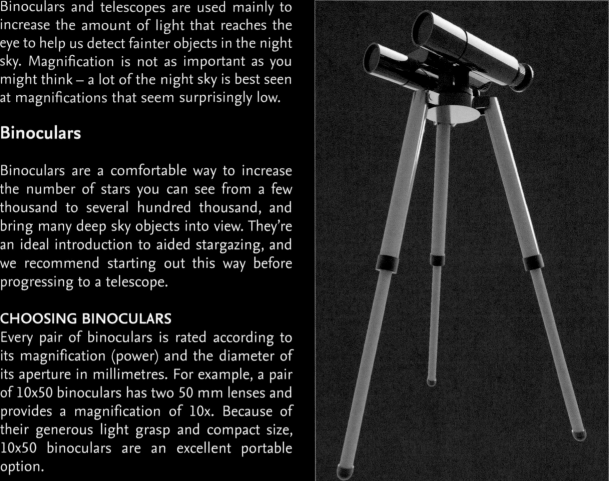

Binoculars on a tripod mount.

L-bracket to further improve the comfort of viewing for long periods.

WHAT CAN YOU SEE?
Many deep sky objects can be seen with a pair of 10x50 binoculars. This size will reveal stars down to about magnitude 10 and hundreds of extended objects, including nebulae, globular clusters and brighter galaxies. Star clusters such as M44 (The Beehive) and M45 (The Pleiades) are resolved beautifully even at low magnification, revealing many more stars than you would see with the eye alone.

Binoculars give you a much more interesting view of the Moon, including its dark, smooth plains and rugged highlands, as well as highlighting its most prominent craters

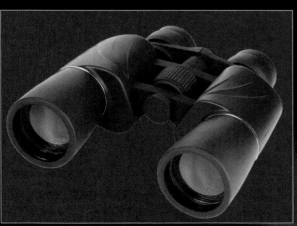

Binoculars, specification 10x50.

Jupiter and the Galilean moons as they appear through a small telescope. Held steady, binoculars also reveal the moons.

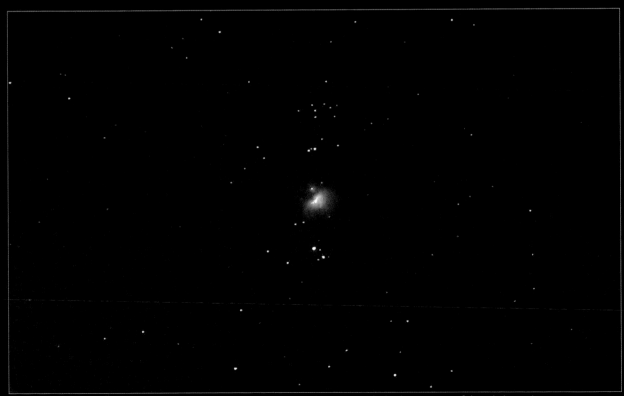

The Great Nebula in Orion (M42) as it appears through binoculars in good conditions. The brightest regions of the nebula can be clearly seen surrounding a cluster of young stars.

USING BINOCULARS OR A TELESCOPE **85**

and ray systems. Looking further out, each planet in our solar system is visible from binoculars although not in great detail. The phases of Venus are revealed, like those of the Moon; Jupiter appears as a round disc flanked by four lights – the large Galilean moons; Saturn may appear slightly elongated at 20 and can be seen with its bright moon Titan and Mercury, Mars, Uranus and Neptune appear as points, and the last three even show some colour.

Each year, several comets can be seen with binoculars, particularly when they are at their brightest. With binoculars, these comets often show extended features, such as tails, and might even have a pronounced green colour produced by fluorescing gases in the coma around the rocky and icy nucleus.

Telescopes

Telescopes have been used for centuries to study the heavens in unprecedented detail. The first telescopes to be used in the early years of the seventeenth century were far less powerful than a modern pair of binoculars. Today, you can buy a very capable telescope for a reasonable price.

CHOOSING TELESCOPES, EYEPIECES AND ACCESSORIES

Telescopes are available in a wide range of sizes and designs. The best telescope is the one you use the most, and it is not necessarily the biggest. There are four main telescope designs commonly found on the market today: refractors, Newtonian reflectors, Schmidt-Cassegrain reflectors and Maksutov-Cassegrain reflectors. Refractors have a lens as the primary light collector and reflectors have a mirror. To get advice on choosing a telescope that best suits your needs you could contact or join a local astronomy club whose members will be happy to help. Telescopes are precisional optical instruments and guidance is highly

recommended to ensure you are comfortable using your new telescope and that you get the best experience.

An eyepiece is used in all telescopes to capture a small part of the centre of the image and blow it up in size before it is projected into your eye. Generally speaking, a telescope should not be used at a magnification that is greater than twice its own aperture in millimetres. For example, a 130 mm (5.12 inches) aperture telescope

A reflecting telescope (the light-gathering component is a mirror). The image is seen from the front or side of the telescope. This Newtonian reflector is one of several reflector designs.

should not exceed 260x magnification. A greater magnification would produce a fainter image with no extra detail. You can calculate the magnification you'll get by dividing the focal length of the telescope by the focal length of the eyepiece. For example, a telescope with a focal length of 1500 mm (59 inches) combined with an eyepiece of focal length 25 mm (1 inch) will produce an effective magnification of 60x. Barlow lenses are an inexpensive way of doubling your eyepiece collection, with some offering a 3x amplification factor.

There are various filters available such as contrast filters, Moon filters, light pollution reduction filters and colour filters that isolate features on different planets. Once you are more familiar with a telescope, you can explore these filters according to your particular interests.

Many telescopes include computerized software called GoTo systems. When properly aligned, computerized telescopes offer very precise tracking. In combination with an equatorial mount, a GoTo system becomes an excellent imaging platform, eventually allowing you to explore deep sky telescopic astrophotography.

WHAT CAN YOU SEE?
Because of their flexibility, telescopes are the most versatile way to explore the sky.

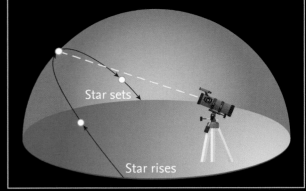

An equatorial mount allows a telescope to track the sky.

Wide-field, low-power views are ideal for spotting star clusters, nebulae, globular clusters and galaxies. Keeping the power low will result in a brighter image for your eye, making a faint target more easily visible. At high power, telescopes are the essential way to tour the Solar System. Telescopes bring our dynamic solar system into sharp relief.

On most nights, the practical limit of magnification is imposed by the transparency and steadiness of the atmosphere to around 250x. Exceptional nights will allow you to exceed 300x, but even on an average night a great number of details can be seen, such as the large Cassini Division in Saturn's rings, the Great Red Spot on Jupiter or the bright polar caps of Mars.

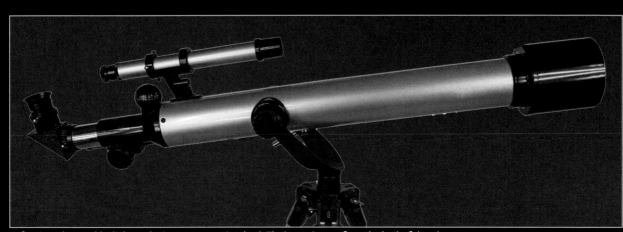

Refracting telescope (the light-gathering component is a lens). The image is seen from the back of the telescope.

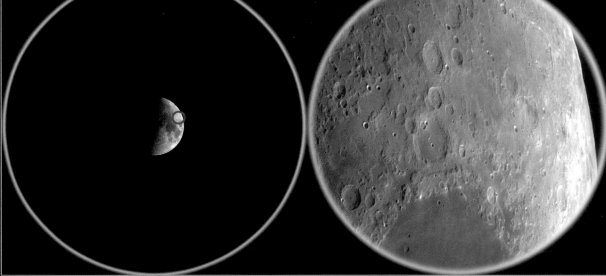

The Moon as seen with the unaided eye, through large binoculars or a low power telescope,

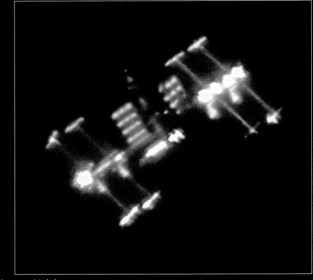

The International Space Station as seen through binoculars (left) and a telescope (right).

Saturn photographed through the eyepiece of a small telescope. The rings are clearly seen encircling the planet.

THINGS TO SEE

Exploring the Solar System

Whether you're using a telescope or not, our solar system provides many observing opportunities.

THE MOON

Earth's Moon is the second brightest object in the sky, after the Sun. During the Moon's 29.5-day synodic month, the phases are said to wax on and wane off. The phases are described below. The number of days given are averages, and vary slightly due to the Moon's elliptical orbit.

Waxing Crescent: 0–7.4 days. The Moon's eastern edge (the limb) emerges (east on the Moon is west in the sky). It sets in the early evening.

First Quarter: 7.4 days. Exactly half of the Moon's near side appears to be illuminated. The boundary between the bright and dark sides is called the terminator.

Waxing Gibbous: 7.4–14.8 days. Illumination is greater than 50 per cent, but less than 100 per cent. Its western hemisphere is gradually revealed.

Full Moon: 14.8 days. At 100 per cent phase illumination, the day side of the Moon is directly pointed at the night side of Earth. Since the terminator is not visible, there is very little sense of relief on the lunar surface.

Waning Gibbous: 14.8–22.2 days. The Moon's terminator emerges on the eastern limb.

Last Quarter: 22.2 days. Only the western hemisphere of the Moon appears to be illuminated. As the Moon is now leading the Sun, it rises in the early morning.

Waning Crescent: 22.2–29.53 days. The terminator moves westward towards the western limb (east in the sky).

New Moon: 29.53/0 days. At the moment of New Moon, a new lunar synodic month begins. The New Moon is the night side of the Moon, and too dark to see through the glare of the Sun. New Moon is only visible during a solar eclipse.

In the crescent and quarter phases you may be able to see the rest of the Moon faintly lit – this is called Earthshine and is due to light reflecting off the surface of the Earth onto the Moon.

Features on the Moon fall into one of several categories: craters (circular depressions usually formed from impacts); maria (seas, large basins of solidified lava); montes (large chains of mountain ranges formed by gigantic asteroid impacts billions of years ago); mons (individual mountains); valles (valleys formed by lava flows and collapsed lava tubes); and sinus (rugged bays that border smooth plains with shallow circular depressions).

Due to atmospheric seeing, very small objects, such as those left on the Moon by the Apollo astronauts, cannot be resolved through Earth's atmosphere, but they have been imaged using the Lunar Reconnaissance Orbiter, which was only a few kilometres above the lunar surface.

THE PLANETS

To the unaided eye, the planets appear as steady points of light in the sky. Mercury, Venus, Mars, Jupiter, Saturn and Uranus are visible to the unaided eye. Neptune falls below the limit of what the unaided eye can detect. Use binoculars on a tripod to see the discs of Venus and Jupiter, as well as Jupiter's four brightest moons.

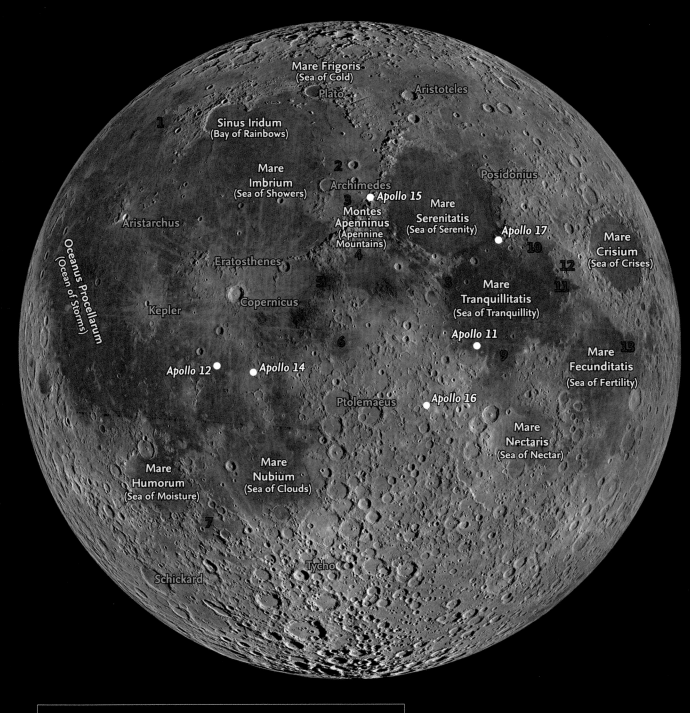

Mare Frigoris
(Sea of Cold)

Plato

Aristoteles

1

Sinus Iridum
(Bay of Rainbows)

2

Posidonius

Mare
Imbrium
(Sea of Showers)

Archimedes

3 ● Apollo 15

Mare
Serenitatis
(Sea of Serenity)

Apollo 17
●

Mare
Crisium
(Sea of Crises)

Aristarchus

Montes
Apenninus
(Apennine
Mountains)

10

4

12

Eratosthenes

Oceanus Procellarum
(Ocean of Storms)

5

Copernicus

8

Mare
Tranquillitatis
(Sea of Tranquillity)

11

Kepler

6

Apollo 11
●

Apollo 12
●

Apollo 14
●

9

Mare
Fecunditatis
(Sea of Fertility)

13

Ptolemaeus

Apollo 16
●

Mare
Nectaris
(Sea of Nectar)

Mare
Humorum
(Sea of Moisture)

Mare
Nubium
(Sea of Clouds)

7

Schickard

Tycho

Bays and Marshes

1	Sinus Roris	9	Sinus Asperitatis
2	Sinus Lunicus	10	Sinus Amoris
3	Palus Putredinis	11	Sinus Concordiae
4	Sinus Fidel	12	Palus Somni
5	Sinus Aestuum	13	Sinus Successus
6	Sinus Medii		
7	Palus Epidemiarum	●	Apollo landing sites
8	Sinus Honoris		

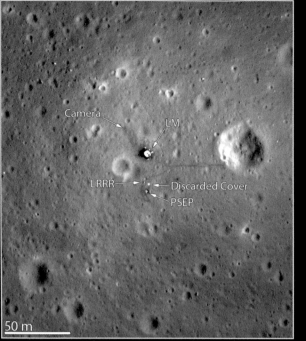

The Apollo 11 landing site in the Sea of Tranquility. Image taken by NASA's Lunar Reconnaissance Orbiter from a height of 24 km (15 miles).

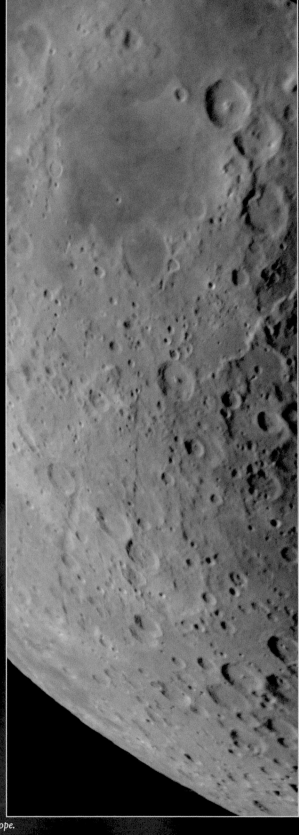

Left: Close up of lunar craters. Right: The Moon seen through a large telescope.

Mercury

The innermost planet Mercury is a small, rocky world about two-fifths of the diameter of Earth, separated from us by an average of 77 million km (48 million miles). With high power, it's possible to resolve Mercury into a disc and observe its phases. Mercury is also relatively close to the Sun, and its maximum angular distance from the Sun of 28° makes it appear low down in the evening or morning sky. Wait until the Sun is completely below the horizon when viewing Mercury through binoculars or a telescope, to avoid the risk of damaging your eyes.

Venus

Venus is the brightest planet as seen from Earth, with a maximum apparent magnitude of -4.9 when it is in a crescent phase. Unlike the Moon, Venus is brighter in its crescent phase than in its gibbous phase, because it is considerably closer to us in this phase. The phases of Venus are easily observed with

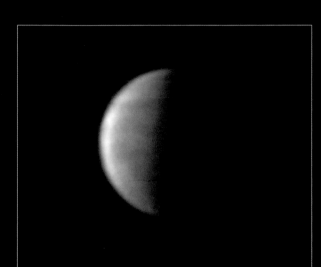

Mercury.

Venus in crescent phase as seen through the Great Equatorial Telescope (71-cm/28-inch aperture), Royal Observatory Greenwich.

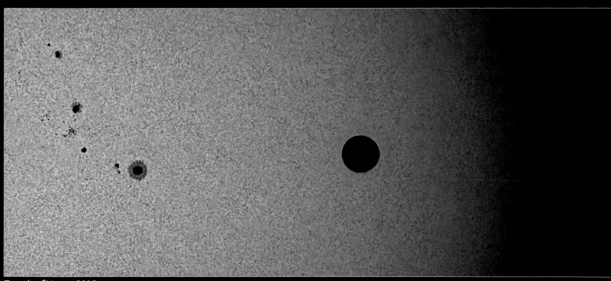

Transit of Venus, 2012.

East ←

West →

Sunset

Sunrise

Path of the Sun if it were visible from the surface of Venus instead of being obscured by the thick atmosphere – the Sun moves in the opposite direction (west to east) compared to on Earth.

binoculars or a telescope, even at lower powers. Its brilliance has earned it the nicknames 'evening star' and 'morning star', depending on whether it is east or west of the Sun. Unlike most of the other planets, including Earth, Venus spins in the opposite direction so that on Venus the Sun rises in the west.

Mars

The red planet is the second nearest to us after Venus. At opposition, when it is closest to us

and visible at night, Mars can be as close as 55 million km (34 million miles).

Although Mars appears quite small, a telescope will reveal its bright icy polar caps and some of the largest markings on its surface, such as Syrtis Major Planum, an exposed outcrop of dark basalt-rich rock. Larger telescopes will reveal finer details and even cloud formations from Mars' major mountains and basins.

Jupiter

The Solar System's largest planet is an ever-changing world of swirling clouds. Jupiter is considerably further away than Mars by hundreds of millions of kilometres. However, it's also much larger than any of the rocky planets and shows the greatest range of colour.

Jupiter's equatorial cloud belts are visible even with small telescopes. At any given time you will also see up to four points of light accompanying Jupiter. These are Io, Europa, Ganymede and Callisto, the Galilean moons,

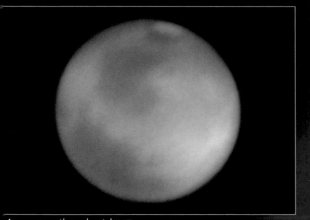

Mars as seen through a telescope

Jupiter and its innermost large moon, Io.

named after the scientist who discovered them. They are the most significant of Jupiter's 67 known satellites. With a telescope, it is possible to see the Galilean moons casting shadows on the surface of Jupiter, known as transit shadows. The Great Red Spot (GRS) is a famous storm in Jupiter's southern hemisphere, embedded within the southern equatorial belt. Its quick rotation period (under 10 hours), means you can see the GRS when Jupiter is near opposition.

Saturn

Saturn is a gas giant nearly on the scale of Jupiter. Even at its closest, Saturn is almost 1,207 million km (750 million miles) from Earth, but its enormous ring structures are easily observed, even with a small telescope.

Titan

Rhea

Dione

Enceladus

Tethys

Janus

Saturn and its moons.

with powerful binoculars, Saturn may seem to have an elongated shape. With larger telescopes, a dark gap in the rings known as the Cassini Division can be seen. On its leisurely 29.5-year orbit around the Sun, the apparent angle of Saturn's rings changes over time. Saturn's rings are composed of innumerable icy particles, most of which are no larger than a grain of sand, but some of which are hundreds of metres across. These are believed to be remnants of comets, asteroids and moons that have collected and collided around Saturn.

With a large telescope, it's also possible to spot some of the largest of Saturn's 62 known satellites, such as Rhea, Tethys, Dione and Iapetus. The biggest of all its moons is Titan, bright enough to be seen even with binoculars.

Uranus

Uranus is a far off, cold world, with an apparent diameter of just four arcseconds at best. Nevertheless, at around four times the diameter of Earth, it is big enough not only to present a distinct disc, but also a pale pea-green colour. A few of its moons may be seen with very large telescopes. Titania and Oberon are quite apparent, whereas Ariel and Umbriel appear faint within the glare of the planet.

Neptune

The most remote planet in our solar system, Neptune appears to be tiny at all times, but

Name	Type	Distance from the Sun (km/ miles/ Astronomical Units)	Diameter (km/miles/ Earths)	Orbital period	Magnitude at brightest	Max. angular diameter at brightest (arcseconds)
Mercury	Terrestrial	57,910,000 km 36,000,000 miles 0.387 A.U.	4,800 km 2,983 miles 0.383 Earths	88 days	-2.45	13.00
Venus	Terrestrial	108,200,000 km 67,200,000 miles 0.723 A.U.	12,100 km 7,519 miles 0.949 Earths	225 days	-4.89	63.00
Mars	Terrestrial	227,940,000 km 141,600,000 miles 1.524 A.U.	6,800 km 4,226 miles 0.532 Earths	1.88 years	-2.91	25.13
Jupiter	Gas Giant	778,330,000 km 483,600,000 miles 5.203 A.U.	142,800 km 88,736 miles 11.21 Earths	11.86 years	-2.94	50.59
Saturn	Gas Giant	1,424,600,000 km 885,200,000 miles 9.523 A.U.	120,660 km 74,978 miles 9.45 Earths	29.46 years	-0.49	21.37
Uranus	Ice Giant	2,873,550,000 km 1,785,600,000 miles 19.208 A.U.	51,800 km 32,189 miles 4.01 Earths	84.01 years	5.32	4.08
Neptune	Ice Giant	4,501,000,000 km 2,796,900,000 miles 30.087 A.U.	49,500 km 30,759 miles 3.88 Earths	164.8 years	7.78	2.37

Planetary data.

Uranus and its moons as seen through the Great Equatorial Telescope (71-cm/28-inch aperture), Royal Observatory Greenwich.

Neptune and its largest moon Triton.

as with Uranus, it can present a small disc. In good conditions its deep blue appearance is unmistakable. Neptune appears so small and faint that it was discovered indirectly by measuring discrepancies in the speed of Uranus. With a very large telescope, it is just possible to observe the largest of Neptune's 14 known moons, Triton. It appears as a very faint star accompanying the ice giant.

DWARF PLANETS
Pluto: With patience and a keen eye, it's possible to see the dwarf planet Pluto, which appears as a faint magnitude +14 star in the constellation Sagittarius. At least a 200-mm (8-inch) aperture telescope is required to make it bright enough to be spotted.

Ceres: The largest inhabitant of the main asteroid belt, Ceres is so small (950 km or 590 miles in diameter) that its apparent size is always under one arcsecond as seen from Earth. For this reason, Ceres appears simply as a point of light, but it makes a relatively rapid motion against the background stars, so observing it from one night to the next is a great way to see the Solar System in motion.

ASTEROIDS
In the years following the discovery of Ceres in 1801, several new 'planets' turned up, including Vesta, Juno and Pallas. Today we recognize that these objects are asteroids, and large members of the main asteroid belt. Such large asteroids are quite easy to observe with binoculars and telescopes, each appearing simply as a point of light. In fact, the term 'asteroid' literally means 'star-like'.

The table on page 102 lists prominent asteroids, along with their apparent magnitudes, which can be found using binoculars or telescopes.

SPOTTING THE INTERNATIONAL SPACE STATION
The Moon is Earth's largest satellite, but its second largest is entirely artificial. The International Space Station (ISS) is an orbiting laboratory travelling around the entire planet more than 15 times every day. It's staffed by an international crew of astronauts and cosmonauts who perform scientific experiments and record spectacular views of Earth from an altitude of 400 km (250 miles). The ISS is so big that it can reflect enough sunlight to outshine even Venus.

With binoculars or a telescope, it is possible to see the station's major components,

The path of the ISS across the sky over London, UK.

The International Space Station in orbit.

Name	Maximum magnitude	Diameter in km (miles)
Vesta	5.2	529 (329)
Pallas	6.49	544 (338)
Ceres	6.65	952 (592)
Iris	6.73	200 (124)
Eros	6.8	34 (21)
Hebe	7.5	186 (116)
Juno	7.5	233 (145)
Melpomene	7.5	141 (88)
Eunomia	7.9	268 (167)
Flora	7.9	128 (80)

The brightest asteroids to look for.

including the large solar arrays that provide its power.

The best passes (the tracks of the ISS) are bright, long and high above the observer's local horizon. Some passes are relatively low, and some are cut short by the Station moving into Earth's shadow, as if it suddenly disappears. You can find information about upcoming passes of the ISS for anywhere in the world by using NASA's Spot The Station service online. Visit spotthestation.nasa.gov.

COMETS

While planets and minor planets are relatively predictable, comets can arrive at short notice and light up our skies. Every year, a selection of comets can be seen using binoculars and telescopes, with

Comet Hale-Bopp seen in 1997.

A photograph showing how Comet C/2011 14 (PANSTARS) appeared to the naked eye.

A bright Perseid meteor passes near the Andromeda Galaxy (right) away from the radiant in Perseus. The Double Cluster is seen on the left.

the occasional visitor bright enough to be spotted by the unaided eye.

Comets appear as extended objects – unlike stars, which are seen as point sources of light. Despite mostly appearing faint, comets are a wonderful, ever-changing sight. Many comets show a pale or even pronounced green colour when observed through instruments. This is due to the presence of diatomic carbon (C_2) in the gaseous atmosphere (coma) that emerges from the cometary nucleus, which fluoresces green when energized by radiation from the Sun.

When comets are discovered, they are tracked by astronomers so that their peak brightness can be predicted, but comets can surprise us and rapidly brighten without warning.

METEOR SHOWERS
There are many annual meteor showers with predictable peak dates and rates. The table on the next page gives some of the best annual meteor showers.

Radiants for these showers (the area in the sky from which most meteors diverge outwards) are included on the seasonal star charts.

The Zenithal Hourly Rate (ZHR) describes the number of meteors each hour under ideal conditions. You should expect to see fewer than this, depending on your local conditions and the position of the Moon. Some showers have a history of producing storms, where meteors are seen at a rate of several per minute (or even every few seconds!) for a short period of time, but such storms are virtually impossible to predict.

During a shower, meteors can be seen all over the sky, and there's no advantage to using binoculars or telescopes, which restrict your field of view. The best way to observe a shower is to settle into a reclined chair and allow your eyes to relax and focus on the stars. To increase your chances of catching meteors, why not use a DSLR camera on a tripod to take continuous long exposures of an area? Meteors will appear as bright streaks on the image, with one end brighter than the other. Lines of continuous brightness are usually satellite trails, rather than meteors.

ECLIPSES

Lunar eclipses are safe to observe telescopically or with binoculars, since all lunar eclipses are fainter than an ordinary Full Moon. A timelapse of photos taken with the same exposure will reveal the gradual change in the Moon's apparent brightness. Due to the size of Earth's shadow at the distance to the Moon, partial and total lunar eclipses are quite long events, so if at first the weather is not cooperative, persevere. You may get a break in the clouds during part of the eclipse.

Watching solar eclipses is a form of solar observing, so proper care must be taken. The safest way to view any solar eclipse

is with a pinhole projector. A tiny hole in the centre of a piece of card, when held up to the Sun, will project an image of the solar disc in its shadow. You can easily construct a pinhole projector from a breakfast cereal box by pricking a small hole in one side (at the top), holding it up to the Sun and looking inside to see the image. In fact, any small hole will act as a pinhole projector – you can even use a colander to safely produce hundreds of tiny eclipse images!

You can observe solar eclipses directly only with correctly filtered optics or dedicated eclipse glasses. Do not attempt to watch an eclipse using sunglasses, welding goggles or anything else not designed for the specific purpose of solar viewing. Telescopes can be filtered at the front using Baader Planetarium's AstroSolar Safety Film. The entire clear aperture must be covered with the visual (ND5.0) version of the film or permanent eye damage will occur.

Shower	Dates of activity	Maximum possible hourly rate
Quadrantids	January 1–10	120
Lyrids	April 16–25	18
Eta Aquarids	April 19 to May 26	55
Alpha Capricornids	July 11 to August 10	5
Perseids	July 13 to August 26	100
Delta Aquarids	July 21–23	16
Alpha Aurigids	August to October	10
Southern Taurids	September 7 to November 19	5
Northern Taurids	October 19 to December 10	5
Orionids	October 4 to November 14	25
Leonids	November 5–30	15
Geminids	December 4–16	120
Ursids	December 17–23	10

The best meteor showers throughout the year.

A total solar eclipse – the Sun's outer atmosphere or corona is visible.

A partial solar eclipse.

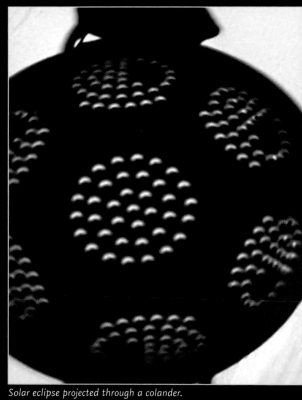

Solar eclipse projected through a colander.

Left: A total lunar eclipse. Right: A partial lunar eclipse.

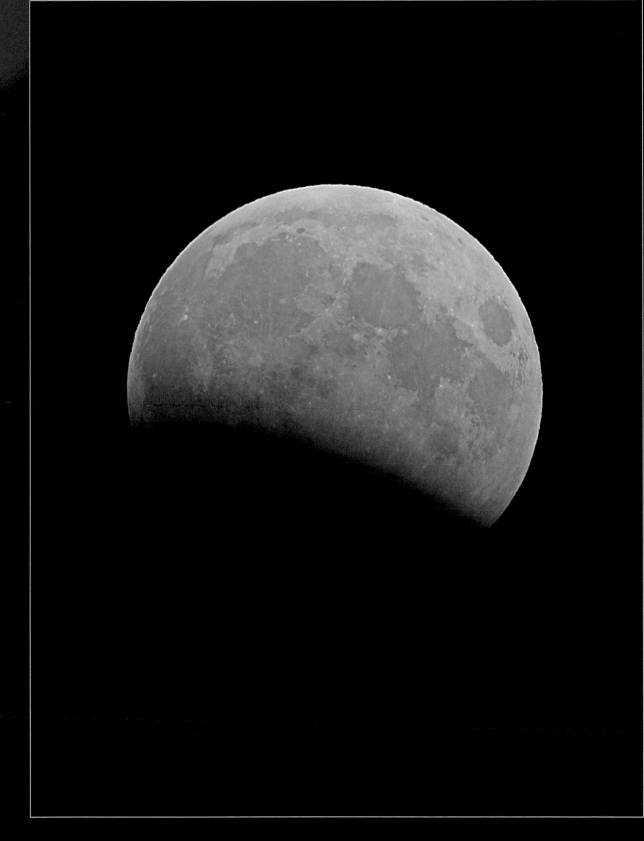

AURORAE

Witnessing a playful, dynamic auroral display requires perseverance, often in cold conditions. There are many good destinations in northern Europe, Canada and Alaska where the *Aurora Borealis*, or northern lights, are visible on most nights of the year. In the southern hemisphere, the southern lights (*Aurora Australis*) can be glimpsed on rare occasions from southern Australia and New Zealand, or better still, Antarctica!

From space, the auroral ovals can be seen encircling Earth's magnetic poles. These ovals never disappear, but the aurorae are too faint to be seen against the blue daytime sky. The best times of year to see the lights are before the local spring equinox (February and March in the northern hemisphere; August and September in the southern hemisphere) and after the local autumn equinox (October and November in the northern hemisphere; April and May in the southern hemisphere).

Even more impressive are geomagnetic storms, which occur when Earth's magnetic field undergoes tumultuous interactions with that of the Sun. In this case, intense bursts of auroral activity, rarely lasting longer than a few hours, light up the skies with a beautiful palette of greens, pinks, violets and reds.

Because the eye's sensitivity to colour is lower when viewing the night sky, the colours of the aurorae appear far less vivid than they do in photographs. The images below show the view as recorded by a camera, with the same scene as it would appear to the dark-adapted eye. However, the colours are easy to detect by eye, when the display is bright and fast-moving, even if only for fleeting moments.

Photograph of Aurora Borealis taken from Iceland (left), approximate naked eye appearance (right).

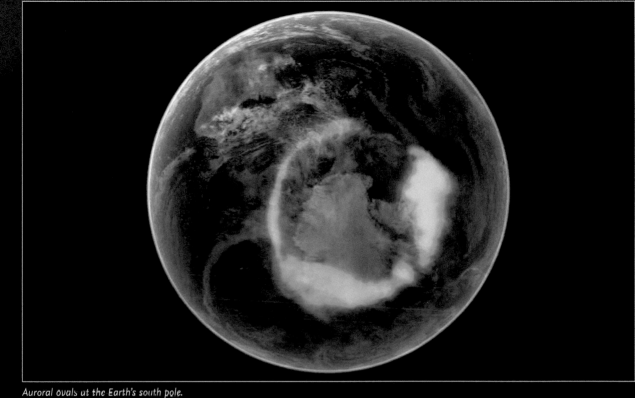
Auroral ovals at the Earth's south pole.

Our Milky Way is a huge collection of stars – possibly up to 400 billion – in a vast magnificent spiral structure. There are 60 times more stars in our galaxy than there are people living on Earth. In fact, throughout the observable Universe, there are at least 10,000 stars for every grain of sand on all of Earth's beaches, and likely more potentially habitable worlds than there have been heartbeats in every human being that has ever lived. Light takes 150,000 years to traverse our galaxy. Our whole solar system is embedded in one of the spiral arms called Orion, about 26,000 light-years from the galactic centre, which lies in the constellation Sagittarius. Right in the heart of our galaxy lies a supermassive black hole – we know this exists because of the strong gravitational effects it has on nearby stars, pulling them round into tight orbits.

On a dark clear night away from city lights you should be able to see around 5,000 stars, and with binoculars this number shoots up to tens and even hundreds of thousands. The bigger the aperture of your telescope or binoculars, the more light it can collect and the deeper you can probe into the cosmos. The limiting magnitude of the eye is about +6 if the skies are dark and clear. In the city, the faintest objects you can see have an apparent magnitude of around +3 to +4.

STAR FORMING NEBULAE

Later in this book there is a chapter covering constellations where you can find a selection of objects to look for across the year. Most of them are stars at various points in their lifecycles. Stars are formed in clouds of gas and dust – star factories have associated emission nebulae – these are colourful clouds of gas often tens of thousands of degrees hot. High-energy ultraviolet light from young stars excites the different atoms, causing them to release visible light. Red and pink regions reveal hydrogen and nitrogen gas, while a blue-green glow is emitted by oxygen gas. Really hot stars appear blue to us, while red stars are much cooler.

DYING STARS

Once they've exhausted their hydrogen supply, stars reach the next stage in their lifecycle, expanding into a red giant. This will happen to the Sun in about 5 billion years' time, as it is currently halfway through its life. Red giant stars undergo pulsations and these affect their brightness – many of them become variable stars, where their magnitude fluctuates. They will slowly shed their outer layers over hundreds of thousands of years and leave behind a compact white dwarf star. Much larger stars become huge red supergiant stars that would engulf an area almost as large as our solar system, and they eventually explode as a supernova. These explosions are so bright that they outshine the host galaxy, making them easy to spot through small telescopes. On average, two supernovae occur every century in the Milky Way.

MULTIPLE AND VARIABLE STARS

Around 85 per cent of stars in the Milky Way are locked in binary or multiple star systems. Many multiple stars usually appear as double stars. Some of these systems orbit so close together they can't be seen clearly in a telescope – instead, astronomers analyze the light from these stars to see if it has been affected by their motion. Lots of binaries can be seen through a telescope and often the two stars are quite different from each other and may have different colours – these are called binary double stars.

STAR CLUSTERS

Within the Milky Way, stars form together in galactic neighbourhoods – their strong winds blow away the remaining dust and gas and they form open clusters visible to us. These loose stellar groups have up to a thousand members and there are some that can be seen with the naked eye. Globular clusters are compact, densely-packed groups of stars that live in the halo of the Milky Way, outside the flat disc. Some of these clusters, containing mostly older, redder stars, are almost as old as the Universe.

GALAXIES

Beyond the Milky Way are more galaxies of various shapes and sizes – lots of spiral galaxies along with bulky orange elliptical galaxies (rugby-ball shaped) and irregular galaxies that have lost their structures. Some of these galaxies have active supermassive black holes in their cores: these gravitational monsters produce powerful effects on the surrounding matter, tugging at nearby gas and stars and pulling them into a tight disc, which then funnels matter outwards in two jets, like a galactic lighthouse.

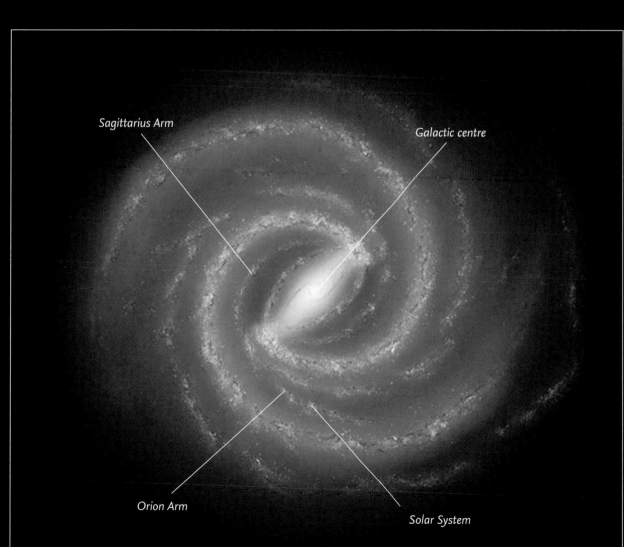

Sagittarius Arm

Galactic centre

Orion Arm

Solar System

Our solar system in the Milky Way.

CONSTELLATIONS AND SEASONAL OBJECTS

There are so many objects to see, photograph and enjoy in our galaxy and beyond, throughout the year and from anywhere on Earth. In this section you will find a selection of constellations to look for from the northern and southern hemispheres and a variety of celestial objects to see with the naked eye, binoculars or a telescope. The night sky is the best show on Earth – go out and look up!

If you are new to astronomy and star charts then you will find useful information and explanations on these pages. There is also a glossary of astronomy terms at the back of the book on pages 216–219. The key shows the symbols used on the chart for stars and other objects. The Greek alphabet will help you understand the importance of the stars in each constellation. They are known as Bayer letters and you will soon recognize them if you use charts or apps regularly.

CONSTELLATION CHARTS

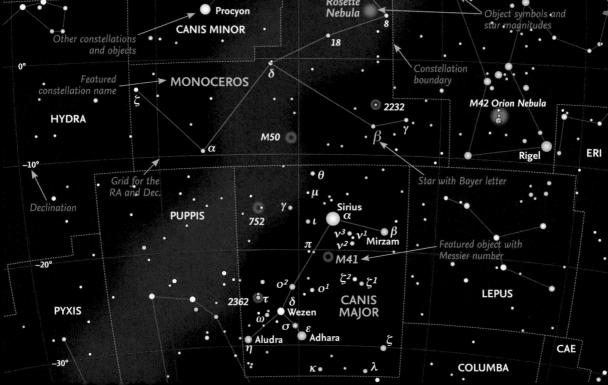

Stellar magnitudes seasonal maps:	● ● ● ● ● • •
	-1 0 1 2 3 4 5
Stellar magnitudes constellation maps:	● ● ● ● • • •

Spectral classes:	● ● ● ● ● ●
	O/B A F G K M
Variable stars:	◉ ◉ ◉ ◉ ◎ ○

OBJECTS

○ ○ Open Star Clusters

● ● Globular Star Clusters

✦ ✦ Planetary Nebulae

☁ ● Bright Nebulae

☁ ☁ Galaxies

Milky Way

Meteor Radiant

The Greek Alphabet

α	Alpha	ε	Epsilon	ι	Iota	ν	Nu	ρ	Rho	φ (φ)	Phi
β	Beta	ζ	Zeta	κ	Kappa	ξ	Xi	σ (ς)	Sigma	χ	Chi
γ	Gamma	η	Eta	λ	Lambda	ο	Omicron	τ	Tau	ψ	Psi
δ	Delta	θ (ϑ)	Theta	μ	Mu	π	Pi	υ	Upsilon	ω	Omega

SEASONAL OBJECTS

We have provided detailed data and information about each object in this section. Below is a brief guide to interpreting these descriptions.

Object name: The name may include a Messier number e.g. M42, an Index Catalogue number e.g. IC 2602 or New General Catalogue number e.g. NGC 246. Star names include their Bayer or Flamsteed designation. Some stars do not have common names, but are listed with a number. These are Flamsteed designations.

TYPE: This may be a nebula, star cluster, galaxy or a star. You can recognize them on the charts by referring to the chart key.

APPARENT MAGNITUDE: A number that measures the brightness of the object as seen by someone on Earth. The brighter an object appears, the lower the magnitude value is.

ANGULAR SIZE: An angular measurement in degrees that describes how large a spherical object appears from Earth. Individual stars do not include this measurement.

POSITION: Right Ascension and Declination are explained on page 24, see also the labels on the chart on the left. On the charts RA labels are in hours. Declination is shown in degrees. Like a map of the Earth, these are positive where they are above the equator or negative if they are below the equator. The charts in this book are ordered to run roughly from the north of the northern hemisphere to the south of the southern hemisphere.

DISTANCE: How far the object is from Earth. One light-year is the distance that light travels in a vacuum in one Earth year of 365.25 days. That is 9,500,000,000,000 km (6,000,000,000,000 miles).

Two titans of the northern sky, Ursa Major (Great Bear) and Draco (Dragon) appear barren, but contain many interesting stars and some oustanding deep sky objects. Ursa Major is home to the Plough, or Big Dipper asterism, which makes up the bear's back and tail.

URSA MAJOR

Mizar and Alcor
(Zeta ζ) Ursae Majoris and 80 Ursae Majoris)

TYPE: Multiple Star System

APPARENT MAGNITUDE: +2.27 / +3.99

POSITION: RA: 13h 23m 56s,
Dec: +54° 55' 31"

DISTANCE: 86 light-years / 81.7 light-years

DESCRIPTION: Mizar sits in the handle or head of the famous Plough or Big Dipper asterism. This actually places it in the unusually long tail of Ursa Major. Mizar and its fainter companion Alcor form a wide double star visible to the naked eye – one of the most famous in the sky. A large telescope will reveal Mizar itself to be a close binary, and today it's known to be a quadruple star. Together with Alcor, which itself is a very close binary, the entire system is composed of six stars! Mizar and Alcor are sometimes known as the horse and rider.

Bode's Galaxy and Cigar Galaxy
(M81 and M82)

TYPE: Spiral Galaxy / Starburst Galaxy

APPARENT MAGNITUDE: +6.94 / +8.41

ANGULAR SIZE: 26.9 arcmin / 11.2 arcmin

POSITION: RA: 09h 55m 33s,
Dec: +69° 03' 55"
RA: 09h 55m 52s,
Dec: +69° 40' 47

DISTANCE: 11.8 million light-years /
12 million light-years

DESCRIPTION: A beautiful spiral with an active black hole in its centre that has a mass 70 million times that of the Sun, M81 has been affected by the gravitational attractions of the smaller galaxy M82. Tidal interactions between the pair have induced vigorous star formation in M82.

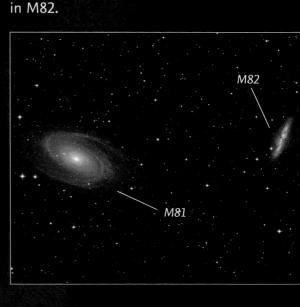

Dubhe

Alcor

Mizar

M82

M81

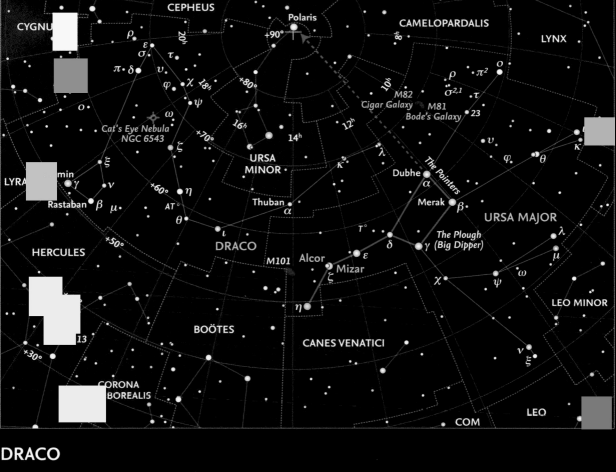

DRACO

Cat's Eye Nebula (NGC 6543)

TYPE: *Planetary Nebula*

APPARENT MAGNITUDE: *+9.80*

ANGULAR SIZE: *20 arcmin*

POSITION: *RA: 17h 58m 33.4s,*
          Dec: +66° 37' 59"

DISTANCE: *3,300 light-years*

DESCRIPTION: This is a wonderful example of a complex planetary nebula – the outflow of a dying star perhaps once similar to the Sun. It has been very well studied, allowing its age to be estimated at about 1,000 years. Its many envelopes of expanding gas and faint blue-green colour are a delight in larger telescopes, and it can be glimpsed with binoculars as a small extended patch of light. This is an object you will come back to many times, as its clarity is very sensitive to atmospheric conditions.

Cassiopeia (Queen) is recognisable as a W or M-shape (depending on the time of year) and harbours a variety of popular deep sky objects. Camelopardalis (Giraffe) is a very barren northern constellation, whose long neck stretches up close to the north celestial pole.

CASSIOPEIA

The Bubble Nebula (NGC 7635)

TYPE: *Emission Nebula*

APPARENT MAGNITUDE: *+10.00*

POSITION: *RA: 23h 20m 48s,*
Dec: +61° 12' 06"

DISTANCE: *7,100–11,000 light-years*

DESCRIPTION: This faint nebula is full of hot hydrogen gas glowing red around a central star larger and hotter than the Sun. The nebula is visible through a large telescope, and a long exposure photograph will reveal some of its colours.

NGC 457

TYPE: *Open Cluster*

APPARENT MAGNITUDE: *+6.40*

ANGULAR SIZE: *13 arcmin*

POSITION: *RA: 01h 19m 33s,*
Dec: +58° 17' 27"

DISTANCE: *7,922 light-years*

DESCRIPTION: In 1787, William Herschel discovered this young star cluster, whose members are no more than 21 million years in age. It contains around 150 stars, and is dominated by two bright members. The pair remind some stargazers of large eyes, so it has been dubbed the Owl Cluster. When you have a clear view of Cassiopeia, it's well worth hunting down this sometimes overlooked object – a jewel fit for a crown.

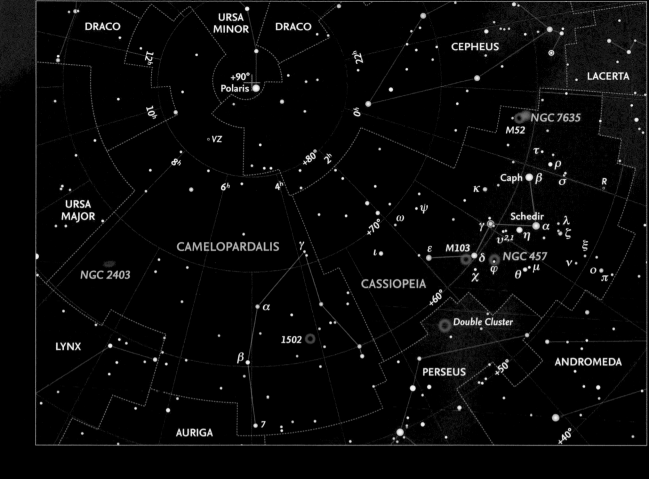

CAMELOPARDALIS

NGC 2403

TYPE: *Spiral Galaxy*

APPARENT MAGNITUDE: *+8.90*

ANGULAR SIZE: *21.9 arcmin*

POSITION: *RA: 07h 36m 51.4s,
Dec: +65° 36' 09"*

DISTANCE: *8 million light-years*

DESCRIPTION: Visible through binoculars, this spiral galaxy is full of clouds of pink hydrogen gas – the key ingredient for star formation. A relatively nearby galaxy, its spiral arms can be resolved using large telescopes. William Herschel discovered this galaxy in 1788.

PERSEUS AND AURIGA

Perseus (Warrior) epitomises the hero's tale. His story has been told for millennia, most recently in the form of 'Clash of the Titans'. Perseus contains the fascinating multiple star Algol (Beta Perseii), commonly called the Demon Star. Auriga (Charioteer) lies near the Milky Way, and is home to several rich open clusters.

PERSEUS

The Double Cluster (NGC 869 and NGC 884)

TYPE: *Open Clusters*

APPARENT MAGNITUDE: *+3.70 / +3.80*

ANGULAR SIZE: *65 arcmin*

POSITION: *RA: 2h 20m,*
Dec: +57° 08'

DISTANCE: *7,500 light-years*

DESCRIPTION: These two clusters can be seen with the naked eye away from city lights. The stars are young, only 13 million years old, some of which are blue-white supergiants. Both clusters are approaching us at incredibly high speeds of around 38 km (24 miles) per second.

APPARENT MAGNITUDE: +0.08

POSITION: RA: 05h 16m 41s,
Dec: +45° 59' 52.8"

DISTANCE: 43 light-years

DESCRIPTION: The bright star Capella comprises two binary systems. The first contains two yellow stars orbiting each other over a period of 104 days. The second contains two red dwarf stars, about 0.16 light-years from the first binary. Capella is a brilliant star, sometimes called the 'Goat Star'. Look nearby to find a trio of fainter stars making up a small asterism called 'The Kids'.

M36 / M37 / M38

TYPE: Open Clusters

APPARENT MAGNITUDE: +6.30 / +6.20 / +7.40

ANGULAR SIZE: 12 arcmin / 24 arcmin / 21 arcmin

POSITION: RA: 05h 36m 12s,
Dec: +34° 08' 04"
RA: 05h 52m 18s,
Dec: +32° 33' 02"
RA: 05h 28m 42s,
Dec: +35° 51' 18"

DISTANCE: 4,100 light-years / 4,511 light-years / 4,200 light-years

DESCRIPTION: A chain of sparkly open clusters is strewn across the constellation Auriga, serving up splashes of light to binocular observers, each inviting a closer inspection with a telescope. Young clusters such as these are a hallmark of our Milky Way, where star formation is an on-going process. Eventually, the trio of clusters will disperse, scattering their stars across the galaxy's spiral arms. Until then, they're excellent targets for a beginner to hunt down and enjoy.

Leo (Lion) is a truly ancient constellation and part of the Zodiac. The stars forming the lion's mane form an asterism known as the Sickle, with Regulus (Alpha Leonis) at the end of the handle. Coma Berenices (Berenice's Hair) is home to many galaxies and has a lovely star cluster to see with the naked eye.

LEO

Algieba (Gamma (γ) Leonis)

TYPE: *Double Star*

APPARENT MAGNITUDE: *+2.08*

POSITION: *RA: 10h 19m 58s,*
Dec: +19° 50' 29"

DISTANCE: *130 light-years*

DESCRIPTION: This popular double star in Leo is about 130 light-years away and comprises K and G-type components with subtle different colours. Due to the brighter K-type star's cooler surface temperature, it has a slightly orange hue, while the G-type companion appears very pale yellow. The two stars take over 500 years to orbit each other.

Leo Triplet (NGC 3628, M65, M66)

TYPE: *Galaxy Group*

APPARENT MAGNITUDE: *+9.4 / +10.3 / +9.7*

ANGULAR SIZE: *15 arcmin /*
8.7 arcmin / 9.1 arcmin

POSITION: *RA: 11h 17m,*
Dec: +13° 25'

DISTANCE: *35 million light-years*

DESCRIPTION: A fine trio of spiral galaxies, the Leo triplet is an excellent opportunity to admire our universe on a grand scale. NGC 3628 (left) joins M65 (top right) and M66 (bottom right). M66 is around 35 million light-years away, but physically close enough to the others to interact with them. NGC 3628 is seen virtually edge on, with a dark dust lane just visible in larger telescopes. Even smaller instruments can reveal all three as faint patches of light.

M65

NGC 3628

M66

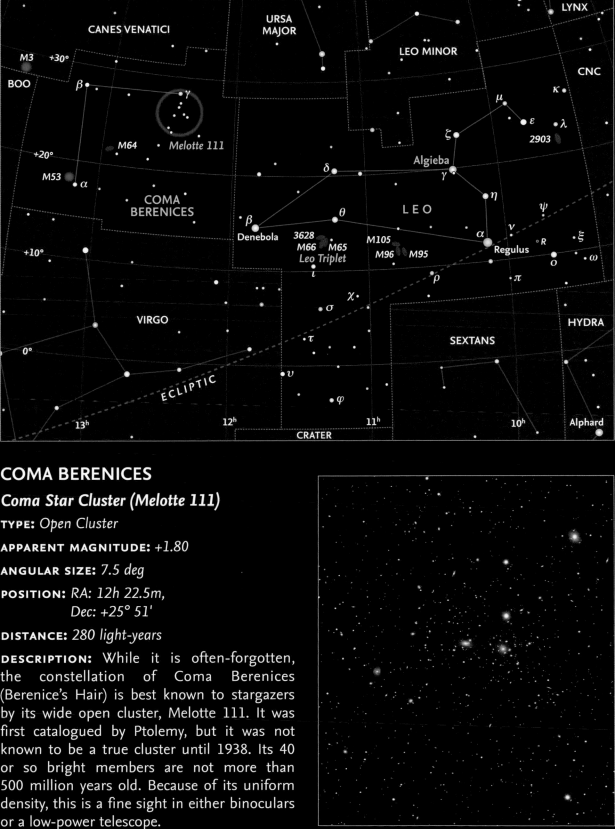

COMA BERENICES

Coma Star Cluster (Melotte 111)

TYPE: *Open Cluster*

APPARENT MAGNITUDE: *+1.80*

ANGULAR SIZE: *7.5 deg*

POSITION: *RA: 12h 22.5m,*
Dec: +25° 51'

DISTANCE: *280 light-years*

DESCRIPTION: While it is often-forgotten, the constellation of Coma Berenices (Berenice's Hair) is best known to stargazers by its wide open cluster, Melotte 111. It was first catalogued by Ptolemy, but it was not known to be a true cluster until 1938. Its 40 or so bright members are not more than 500 million years old. Because of its uniform density, this is a fine sight in either binoculars or a low-power telescope.

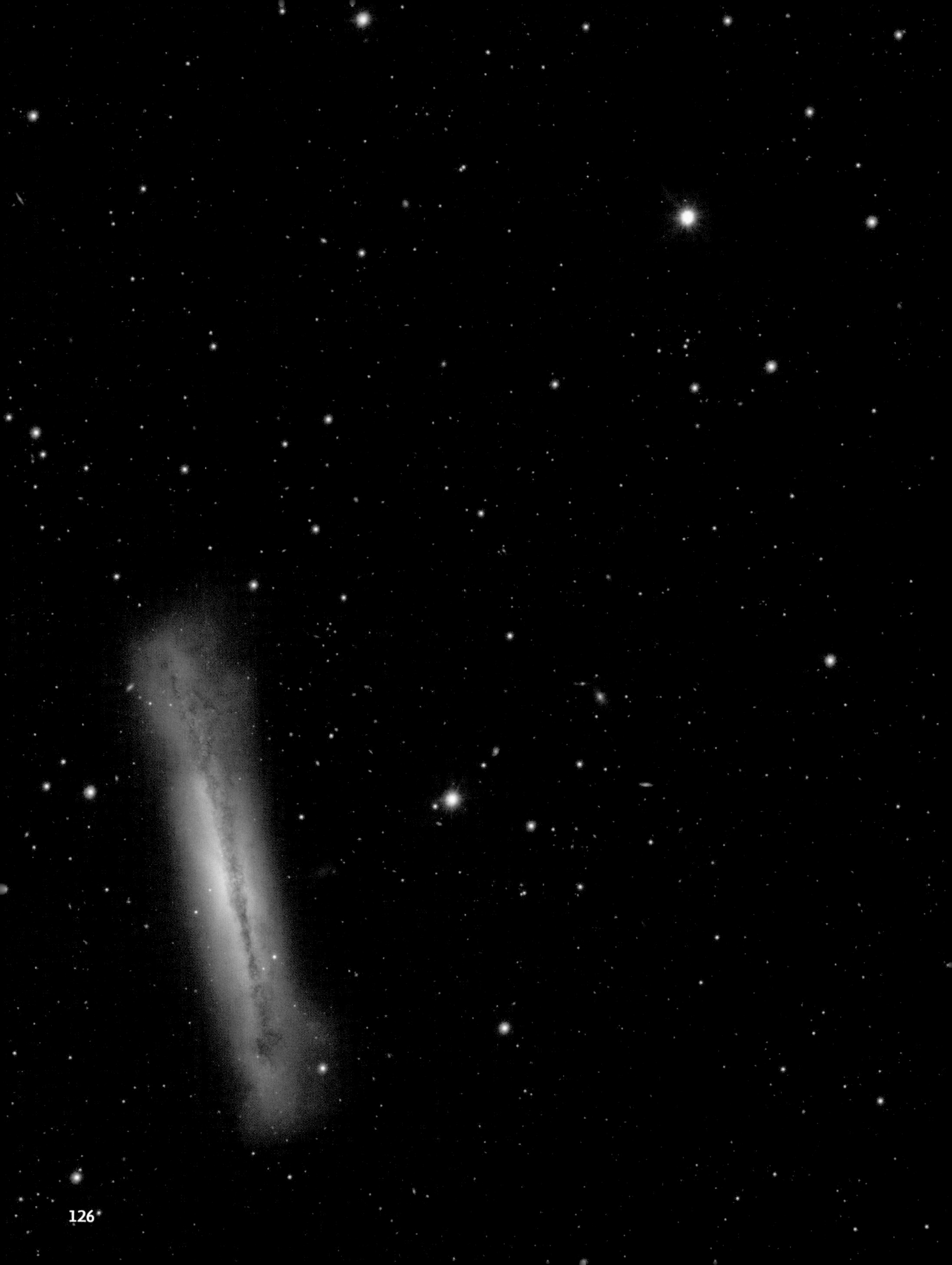

The Leo Triplet in the constellation Leo.

GEMINI AND CANCER

Gemini (Twins) and Cancer (Crab) both lie within the Zodiac. Gemini is known for its two bright stars Castor and Pollux, both of which are of the first magnitude. Castor is a particularly interesting multiple star system. Cancer contains only faint stars.

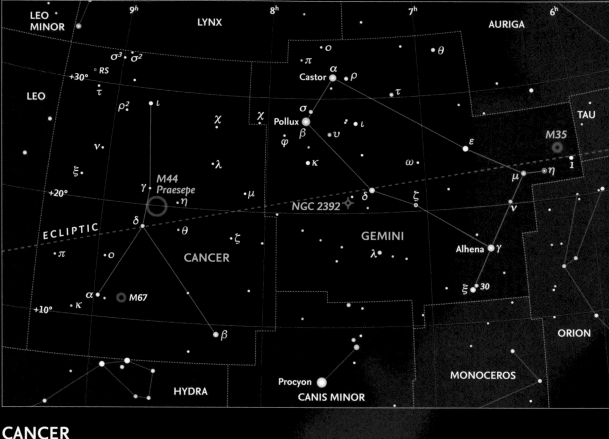

CANCER

Praesepe or Beehive Cluster (M44)

TYPE: Open Cluster

APPARENT MAGNITUDE: +3.70

ANGULAR SIZE: 95 arcmin

POSITION: RA: 08h 40.4m,
Dec: +19° 59'

DISTANCE: 577 light-years

DESCRIPTION: In 1609, Galileo was able to resolve this cluster into 40 stars. It is now known to have 1,000 stars, all gravitationally bound together. There are a mixture of stars at different evolutionary points, including giant main sequence stars, Sun-like stars and older red and white dwarf stars.

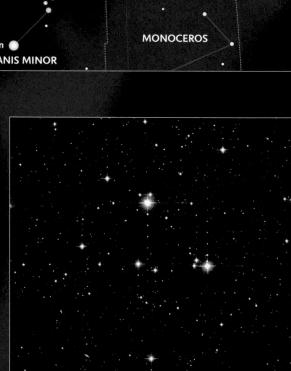

GEMINI

M35

TYPE: *Open Cluster*

APPARENT MAGNITUDE: *+5.30*

ANGULAR SIZE: *28 arcmin*

POSITION: *RA: 06h 08m 54s,*
Dec: +24° 20'

DISTANCE: *2,800 light-years*

DESCRIPTION: To the unaided eye, M35 appears as a speckled patch of light. Viewed at low magnification with binoculars or a telescope, it's a dazzling, nearly circular collection of stars, featuring a few orange giants moving off into old age. Most of the stars in this cluster are around 100 million years old. If you're using a telescope, look out for NGC 2158, a large open cluster that looks nearby but is actually very remote. It appears fainter than M35 but is much older and more massive.

Eskimo Nebula (NGC 2392)

TYPE: *Planetary Nebula*

APPARENT MAGNITUDE: *+10.10*

ANGULAR SIZE: *48 arcsec*

POSITION: *RA: 07h 29m 10.8s,*
Dec: +20° 54' 42.5"

DISTANCE: *≥2,870 light-years*

DESCRIPTION: William Herschel, the astronomer who discovered Uranus, also discovered this compact, distant planetary nebula in 1787, cataloguing it as a nebulous star. He described it as 'a very remarkable phenomenon.' The Eskimo Nebula was once a star like the Sun, and is now an expanding shell of gas surrounded by unusual filaments. At its heart, a tiny white dwarf star remains, illuminating the surrounding cloud with its radiation. In a large telescope and with good conditions, you might see its pale blue–green colour.

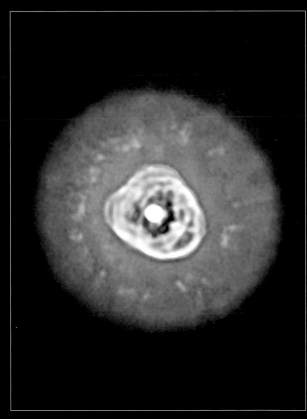

Canis Major (Great Dog) is home to the brightest star in the night sky, Sirius (Alpha Canis Majoris). Monoceros (Unicorn) is relatively faint, but is home to the famous Rosette Nebula.

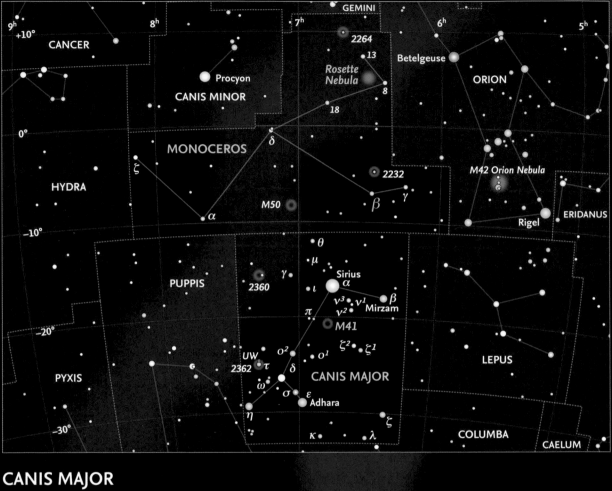

CANIS MAJOR

M41

TYPE: *Open Cluster*

APPARENT MAGNITUDE: *+4.50*

ANGULAR SIZE: *38 arcmin*

POSITION: *RA: 06h 46m,*
Dec: -20° 46'

DISTANCE: *2,300 light-years*

DESCRIPTION: M41 is thought to contain 100 stars, including red giants and white dwarfs. The cluster is about 190 million years old, 25 times younger than the Sun.

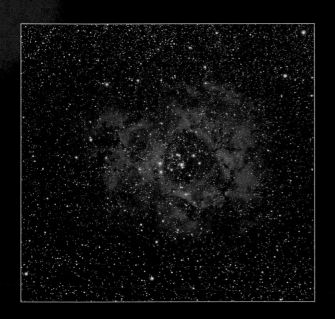

MONOCEROS

Rosette Nebula

TYPE: *Emission Nebula*

APPARENT MAGNITUDE: *+9.00*

ANGULAR SIZE: *1.3 deg*

POSITION: *RA: 06h 33m 45s,*
Dec: +04° 59' 54"

DISTANCE: *5,200 light-years*

DESCRIPTION: This nebula glows red from the energetic hydrogen gas surrounding a cluster of 2,500 blue-white stars, all much hotter and brighter than the Sun. Stars are formed in this huge nebula, 130 light-years wide. At its heart is the cluster NGC 2244, which was discovered by John Flamsteed in 1690.

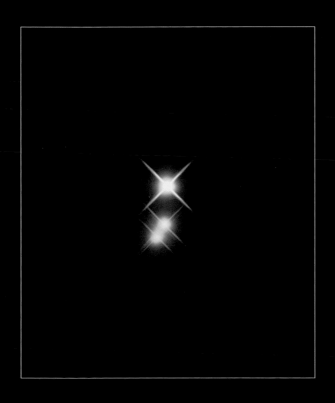

Beta (β) Monocerotis

TYPE: *Multiple Star System*

APPARENT MAGNITUDE: *+3.74*

ANGULAR SIZE: *1.3 deg*

POSITION: *RA: 06h 28m 49s,*
Dec: -07° 01' 59"

DISTANCE: *680 light-years*

DESCRIPTION: To the unaided eye, Beta Monocerotis looks like any other third-magnitude star, but at high magnification it is revealed to be an amazing trio. All three stars are a peculiar type of star that rotates quickly, throwing material into space. They are spinning so fast on their axis that they have become elliptical (rugby-ball shaped). William Herschel described Beta Monocerotis as 'one of the most beautiful sights in the heavens', and we agree!

Orion (Hunter) is one of the most spectacular and recognisable constellations in the sky. Its bright stars include Betelgeuse (Alpha Orionis), a supergiant star poised to end its life in a supernova. Orion is home to a great complex of nebulae, where many stars are forming. Eridanus (River) is a long constellation stretching from Orion's feet to the southern stars.

ORION

The Orion Nebula (M42)

TYPE: *Emission Nebula*

APPARENT MAGNITUDE: *+4.00*

ANGULAR SIZE: *65 arcmin*

POSITION: *RA: 05h 35m 17s,*
Dec: -05° 23' 28"

DISTANCE: *1,344 light-years*

DESCRIPTION: The Orion Nebula can be seen with the naked eye, and the diffuse cloud can be seen through binoculars and a telescope, along with a young open cluster called the Trapezium Cluster. There may be up to 3,000 young stars forming and developing in this giant star factory among the thick clouds of dark dust and cool gas. Images taken with the Hubble Space Telescope show at least 150 protoplanetary discs – these are dusty discs around young stars that will eventually develop into solar systems.

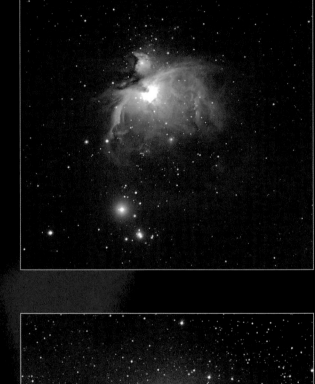

Sigma (σ) Orionis

TYPE: *Multiple Star System*

APPARENT MAGNITUDE: *+3.66*

POSITION: *RA: 05h 38m 45s,*
Dec: -02° 36' 00"

DISTANCE: *1,150 light-years*

DESCRIPTION: Sigma Orionis is actually five stars – the two brightest stars are blue and white, and much hotter, brighter and bigger than the Sun. These two giant stars orbit each other in a relatively close binary, taking 170 years to complete one lap. The other three yellow-white stars orbit this pair at 40 to 100 times further away.

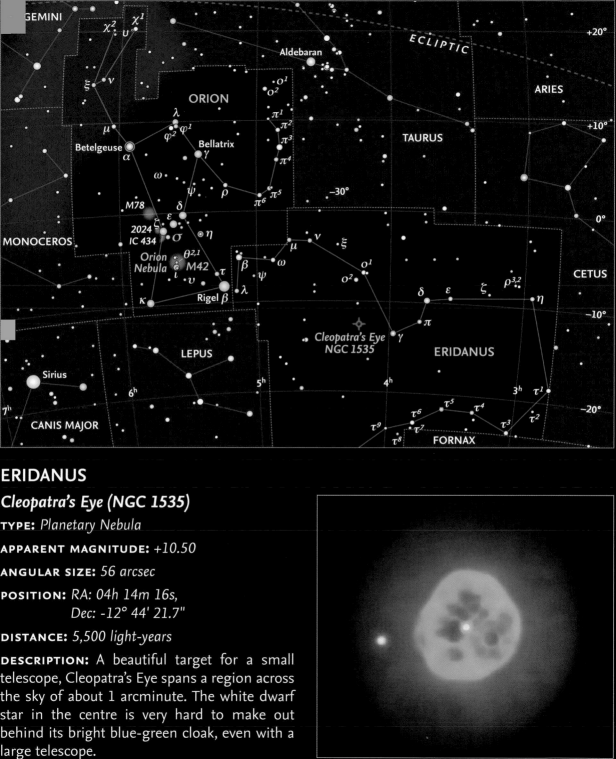

ERIDANUS

Cleopatra's Eye (NGC 1535)

TYPE: *Planetary Nebula*

APPARENT MAGNITUDE: *+10.50*

ANGULAR SIZE: *56 arcsec*

POSITION: *RA: 04h 14m 16s,*
Dec: -12° 44' 21.7"

DISTANCE: *5,500 light-years*

DESCRIPTION: A beautiful target for a small telescope, Cleopatra's Eye spans a region across the sky of about 1 arcminute. The white dwarf star in the centre is very hard to make out behind its bright blue-green cloak, even with a large telescope.

TAURUS

Taurus (Bull) appears to be charging Orion in battle, staring its opponent down with a bright red eye, the giant orange star Aldebaran (Alpha Tauri). This prominent zodiacal constellation is always a welcome sight, thanks to its dazzling pair of nearby star clusters.

TAURUS

Crab Nebula (M1)

TYPE: *Supernova Remnant*

APPARENT MAGNITUDE: *+8.40*

ANGULAR SIZE: *6 arcmin*

POSITION: *RA: 05h 34m 32s,
Dec: +22° 00' 52"*

DISTANCE: *6,500 light-years*

DESCRIPTION: Despite the name, few observers admit to seeing a crab within the tumultuous clouds of this supernova remnant, which has expanded from a supernova explosion recorded by Chinese astronomers in the year 1054. Charles Messier included the Crab Nebula as the first object in his catalogue, denoting it as M1. Binoculars will reveal a faint patch of light, and telescopes are required to tease out its structure. At its heart lurks a pulsar – a neutron star under 32 km (20 miles) wide that rotates around 30 times per second.

The Hyades (Melotte 25)

TYPE: *Open Cluster*

APPARENT MAGNITUDE: *+0.50*

ANGULAR SIZE: *5.5 deg*

POSITION: *RA: 04h 27m,
Dec: +15° 52'*

DISTANCE: *153 light-years*

DESCRIPTION: This prominent star cluster in the head of Taurus the bull is famous for having helped verify Einstein's General Theory of Relativity during the 1919 eclipse observed by British astronomers on an expedition from the Royal Observatory Greenwich. It is the nearest open cluster to the Solar System, and home to the star Ain (Epsilon Tauri) which hosts at least one exoplanet (a planet outside of our solar system) – the first to be discovered in a cluster. At over half a billion years old, the Hyades cluster appears to be dispersing. It's a lovely sight even to the naked eye.

The Pleiades (M45)

TYPE: *Open Cluster*

APPARENT MAGNITUDE: *+1.60*

ANGULAR SIZE: *1.8 deg*

POSITION: *RA: 03h 47m 24s,*
Dec: +24° 07' 00"

DISTANCE: *444 light-years*

DESCRIPTION: Also known as the Seven Sisters, this young cluster can be seen from light-polluted skies. The wispy nebula is a dust cloud that the stars pass through, and their high-energy light cause the surrounding dust particles to glow blue. The cluster itself will disintegrate after about 250 million years, and the ageing stars will migrate to other parts of the galaxy.

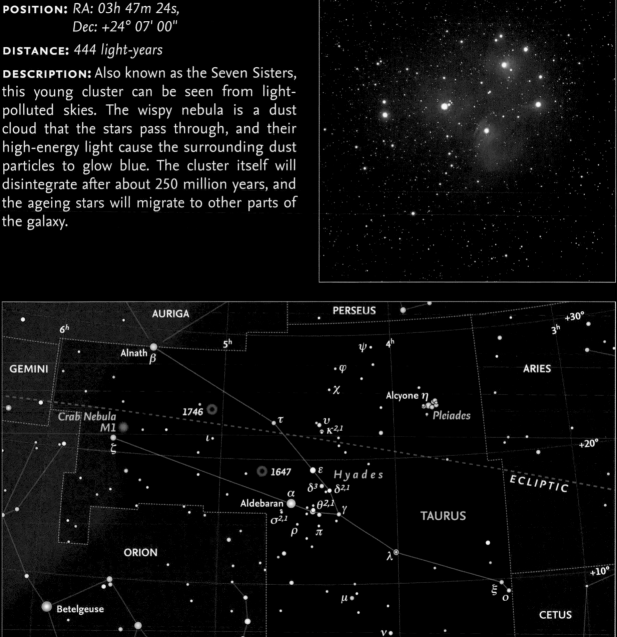

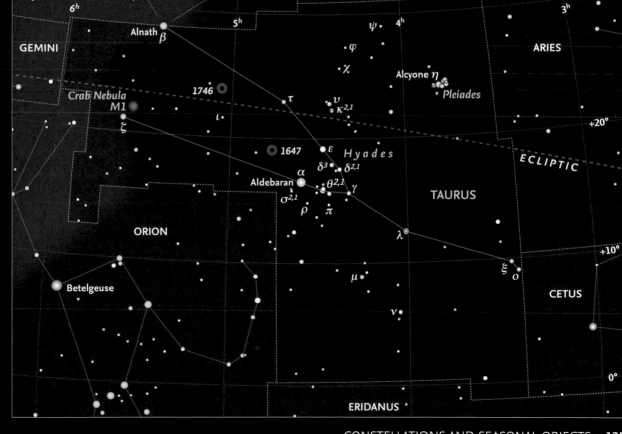

Legend has it that Andromeda (Chained Princess) was sacrificed to a sea monster before being saved by the hero Perseus and his trusty steed Pegasus (Winged Horse). It's here we find the famous Andromeda Galaxy. The body of Pegasus makes up a large asterism called the Great Square.

ANDROMEDA

Andromeda Galaxy (M31)

TYPE: *Spiral Galaxy*

APPARENT MAGNITUDE: *+3.44*

ANGULAR SIZE: *3.2 deg*

POSITION: *RA: 00h 42m 44s,*
Dec: +41° 16' 09"

DISTANCE: *2.54 million light-years*

DESCRIPTION: Our nearest neighbouring large galaxy is a real monster, perhaps twice the size of the Milky Way and home to more than a trillion stars! M31 is a remarkable sight, clearly visible to the unaided eye in dark skies, despite being over 2.5 million light-years away. We see it as it looked 2.5 million years ago. It looks serene, but is on track for a violent collision with our home galaxy in a few billion years. Until then, we can enjoy its bright core, dark dust lanes and large satellites (M32 and M110) which can be seen clearly in a telescope. It's humbling to look back in time to before humans existed and take in an entire galaxy with just your eyes, even if its full extent at nearly six times the width of the Full Moon is too faint to be seen.

Almach (Gamma (γ) Andromedae)

TYPE: *Multiple Star System*

APPARENT MAGNITUDE: *+2.26 / +4.84*

POSITION: *RA: 02h 03m 54s,*
Dec: +42° 19' 47"

DISTANCE: *350 light-years*

DESCRIPTION: For many observers, Gamma Andromedae (Almach) rivals Albireo in beauty. It's a colourful quadruple star system with two visible components of golden orange and azure blue. In fact, the blue star is itself a tight double star with a separation under one arcsecond – extremely challenging to observe! In fact it has another unobservable companion so close it was indirectly detected from the analysis of its spectrum (components of light). The Almach system is approximately 350 light-years away, so its light departed around the time that the Royal Observatory in Greenwich was constructed.

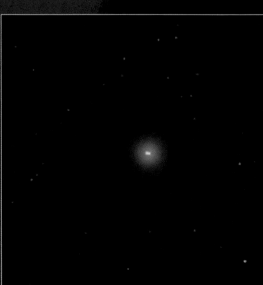

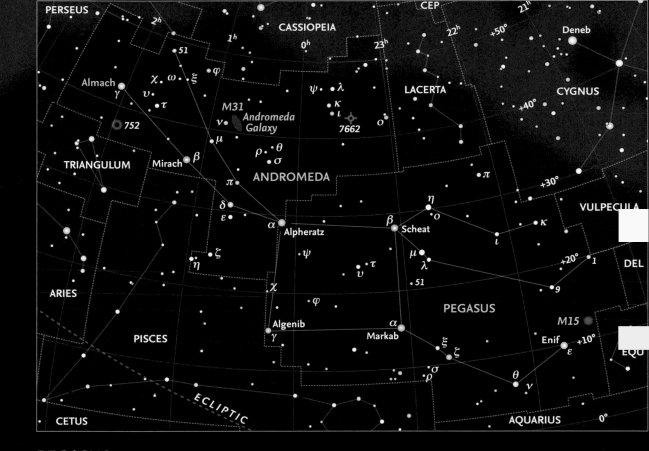

PEGASUS

M15

TYPE: *Globular Cluster*

APPARENT MAGNITUDE: *+6.20*

ANGULAR SIZE: *18 arcmin*

POSITION: *RA: 21h 29m 58s,*
Dec: +12° 10' 01"

DISTANCE: *33,600 light-years*

DESCRIPTION: If there were a carrot dangling in front of Pegasus' nose, it would be M15 – a stunning globular cluster roughly 175 light-years in diameter, home to at least 100,000 stars. M15 is known for its extremely dense core, which might harbour a black hole, and has been well studied in an effort to understand the physics of such clusters. It looks like a fuzzy circular patch in binoculars, but its individual stars become visible with apertures of six inches or larger. Some consider M15 to be an alternative favourite to M13 when comparing northern globular clusters. The night sky from a planet around one of its stars would be so bright from all of the nearby stars that you would be able to read a book.

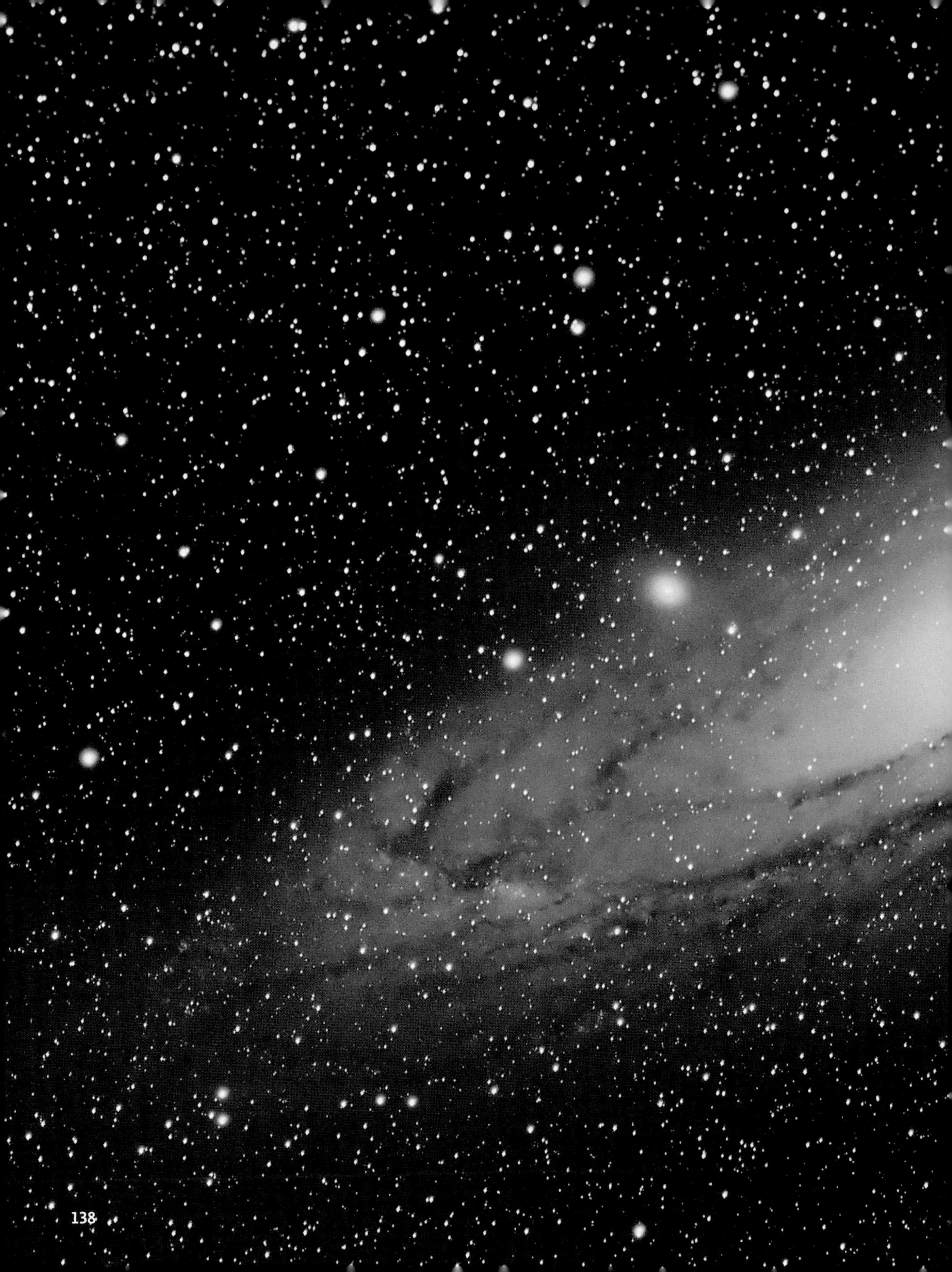

The Andromeda Galaxy (M31) in the constellation Andromeda. **139**

CYGNUS

Cygnus (Swan) is a large constellation containing the well-known asterism the Northern Cross. Cygnus appears to be flying along the Milky Way, and is a rich constellation worth sweeping through with binoculars. Its brightest star Deneb (Alpha Cygni) is a member of the Summer Triangle.

CYGNUS

Albireo (Beta (β) Cygni)

TYPE: *Multiple Star System*

APPARENT MAGNITUDE: *+3.18 / +5.09*

POSITION: *RA: 19h 30m 43s,*
Dec: +27° 57' 35"

DISTANCE: *430 light-years*

DESCRIPTION: The original name of this star, which has since been mistranslated was 'the hen's beak' and it marks the beak of Cygnus the swan. Many stargazers consider Albireo to be the most beautiful pair of stars in the entire sky. Its two components are strikingly different in colour: a powerful azure blue Be-type star (Albireo B) joins a tightly bound pair of stars (Albireo Aa and Ac), which combined, give off a golden hue. The two components of Albireo A are orange and blue, orbiting each other every 213 years. They're too close to be seen separately, so stargazers will have to make do with the wider pairing. Fortunately, it's still a fantastic sight!

Blinking Planetary (NGC 6826)

TYPE: *Planetary Nebula*

APPARENT MAGNITUDE: *+8.80*

ANGULAR SIZE: *27 arcmin*

POSITION: *RA: 19h 44m 48s,*
Dec: +50° 31' 30"

DISTANCE: *2,000 light-years*

DESCRIPTION: This fascinating planetary nebula takes its name from a curious visual phenomenon in which it appears to blink in and out of sight. Though not exclusive to this object, the effect is particularly pronounced in this case. When looking directly at the nebula, the brightness of the central star overwhelms the faint outer shells of gas. If you use averted vision, you'll see the gas pop in and out of view as you move your eye around the field. It's a fun way to see your blind spot at work.

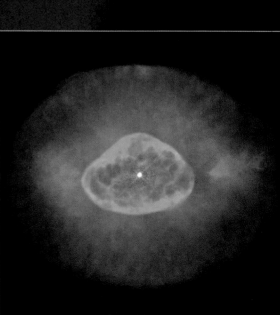

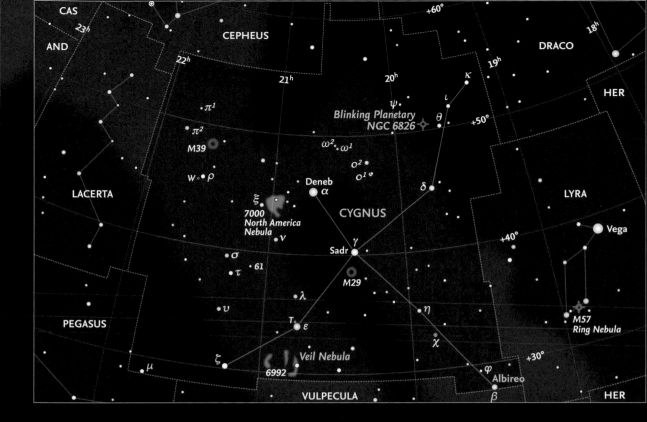

CAS · AND · CEPHEUS · DRACO · HER · LYRA · CYGNUS · LACERTA · PEGASUS · VULPECULA · HER

23ʰ · 22ʰ · 21ʰ · 20ʰ · 19ʰ · 18ʰ
+60° · +50° · +40° · +30°

π¹ · π² · M39 · w · ρ · ξ · 7000 North America Nebula · ν · σ · τ · 61 · υ · λ · ζ · μ · 6992 · τ · ε · Veil Nebula

ω² · ω¹ · o² · o¹ · Deneb · α · Sadr · γ · M29 · η · χ · φ · Albireo · β

Blinking Planetary NGC 6826 · ψ · θ · ι · κ · δ

Vega · M57 Ring Nebula

Veil Nebula

TYPE: *Supernova Remnant*

APPARENT MAGNITUDE: *+7.00*

ANGULAR SIZE: *3 deg*

POSITION: *RA: 20h 42m 38s,*
Dec: +30° 42' 30"

DISTANCE: *1,470 light-years*

DESCRIPTION: We have no record of a massive star in Cygnus ending its life as a brilliant supernova 6,000 years ago, but the evidence that it happened is spread over three degrees of the night sky. A grand example of a supernova remnant, the Veil Nebula is usually divided into several components. The most visible are the eastern and western veils. The western veil (shown) is sometimes called the witch's broom, and appears to surround the star 52 Cygni. An expanding cloud of gas from the supernova heats the surrounding material, causing it to glow. The Veil Nebula is a large, complex object that will keep drawing you back for another look.

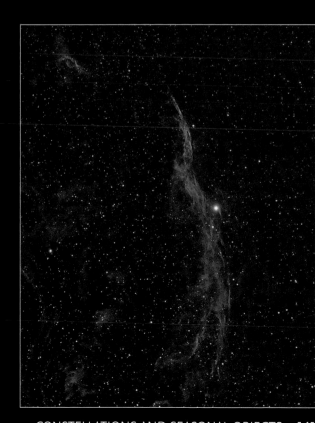

LYRA AND HERCULES

The Legend of Hercules is connected to many other stories in the stars. Hercules (Hero) itself is a sprawling constellation, home to the Keystone asterism. Lyra (Lyre) represents a harp and contains Vega, a star of the Summer Triangle.

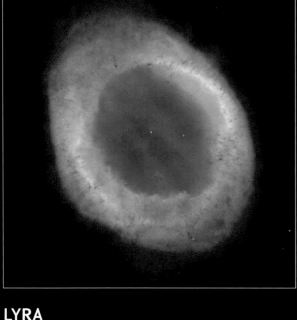

LYRA

Ring Nebula (M57)

TYPE: *Planetary Nebula*

APPARENT MAGNITUDE: *+8.80*

ANGULAR SIZE: *1.4 arcmin*

POSITION: *RA: 18h 53m 35s,*
Dec: +33° 01' 45"

DISTANCE: *2,300 light-years*

DESCRIPTION: When the Sun evolves into its mature, white dwarf stage, it will gradually release shells of gas into the galaxy around it, and might one day look similar to M57 – a planetary nebula that looks like a smoke ring suspended in the sky. The Ring Nebula is really a nearly spherical bubble of gas, at the centre of which we see the remains of a star that began to end its life thousands of years ago.

Epsilon (ε) Lyrae

TYPE: *Multiple Star System*

APPARENT MAGNITUDE: *+4.70 / +5.10*

POSITION: *RA: 18h 44m 20s,*
Dec: +39° 40' 12"

DISTANCE: *162 light-years*

DESCRIPTION: Through binoculars, two components appear separated by several arcminutes. On closer inspection with a telescope, you'll see that each star is in fact a tight double itself, earning Epsilon Lyrae the nickname 'Double Double'. The stars in the individual pairs are separated by just over two arcseconds each, roughly 90 times closer than the separation between the two systems. All four stars are approximately 162 light-years away. This beautiful system is a good test of your telescope and the conditions of the atmosphere.

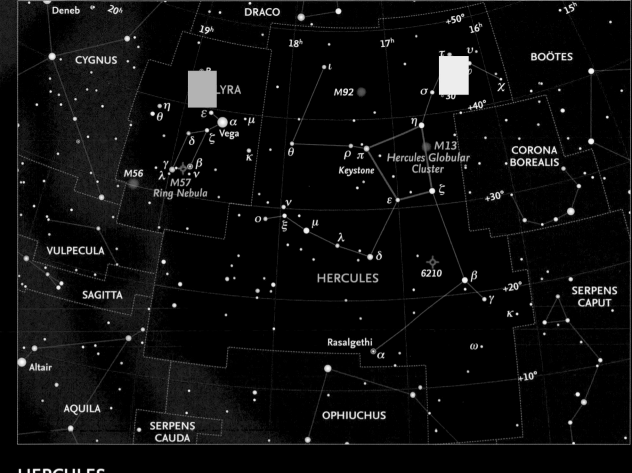

HERCULES

Hercules Globular Cluster (M13)

TYPE: *Globular Cluster*

APPARENT MAGNITUDE: *+5.80*

ANGULAR SIZE: *20 arcmin*

POSITION: *RA: 16h 41m 41s,*
Dec: +36° 27' 36"

DISTANCE: *22,200 light-years*

DESCRIPTION: The great globular cluster in Hercules is one of the best in the sky. Its discoverer Edmund Halley noted: 'It shows itself to the naked eye when the sky is serene and the Moon absent.' A faint patch of light to the eye, M13 comes to life through binoculars and telescopes and is a must-see for any stargazer. Though it cannot quite match the splendour of Omega Centauri, many agree that M13 is the great globular cluster north of the celestial equator. Its stars, in some cases, are estimated to be almost 13 billion years old.

VIRGO AND CANES VENATICI

The large zodiacal constellation Virgo (Maiden) represents Ceres, goddess of agriculture, and is home to a large number of galaxies. They belong to the Virgo Supercluster, along with the Milky Way. Canes Venatici (Hunting Dogs) borders Boötes (Herdsman), like two celestial canines ready to follow their master on a hunt.

CANES VENATICI

Whirlpool Galaxy (M51)

TYPE: *Spiral Galaxy*

APPARENT MAGNITUDE: *+8.40*

ANGULAR SIZE: *11.2 arcmin*

POSITION: *RA: 13h 29m 53s,*
Dec: +47° 11' 43"

DISTANCE: *23 million light-years*

DESCRIPTION: A spectacular face-on galaxy, M51 interacts with a smaller galaxy called NGC 5195 that passed through the main disc about 500 million years ago. M51 has an active black hole in its centre, unlike the Milky Way, whose black hole was once active but is now dormant.

M3

TYPE: *Globular Cluster*

APPARENT MAGNITUDE: *+6.20*

ANGULAR SIZE: *18 arcmin*

POSITION: *RA: 13h 42m 12s,*
Dec: +28° 22' 38"

DISTANCE: *33,900 light-years*

DESCRIPTION: A very fine globular cluster, M3 can easily be seen with moderate telescopes, but comes alive with larger instruments. It appears to have a compact, bright core, but is in fact a large globular cluster a great distance from us, made up of around 500,000 stars. It is further away from us than the centre of our own galaxy: the centre of the galaxy is 26,000 light-years away, but there are globular clusters (extragalactic objects) within 10,000 light-years. M3 is well known to astronomers for having an unusually large number of 'Blue Stragglers' – mature stars that look young, but have actually lost their cooler outer layers.

VIRGO

Sombrero Galaxy (M104)

TYPE: *Spiral Galaxy*

APPARENT MAGNITUDE: *+8.98*

ANGULAR SIZE: *8.7 arcmin*

POSITION: *RA: 12h 39m 59s,*
Dec: -11° 37' 23"

DISTANCE: *28 million light-years*

DESCRIPTION: The Sombrero Galaxy, named for its resemblance to a broad-brimmed hat, is a compact spiral galaxy seen almost edge-on from Earth. It resides in the Virgo supercluster, and is 28 million light-years away. Its unusually bright core is complemented by a dark dust lane of material seen in silhouette. The high contrast of these two features gives this galaxy a striking appearance, and it's a firm favourite among deep sky observers. At magnitude 9, the Sombrero Galaxy is also readily seen with binoculars.

Cetus (Sea Monster) is connected to the story of Perseus and Andromeda, a large constellation of a few bright stars and home to the interesting variable Mira. The faint Sculptor (Sculpture) is home to a few galaxies.

CETUS

Skull Nebula (NGC 246)

TYPE: Planetary Nebula

APPARENT MAGNITUDE: +8.00

ANGULAR SIZE: 3.8 arcmin

POSITION: RA: 00h 47m 03s,
Dec: -11° 52' 19"

DISTANCE: 1,600 light-years

DESCRIPTION: The spooky Skull Nebula was discovered by William Herschel in 1785. As with many planetary nebulae, you can spot its central star, a white dwarf of magnitude +11.8. The nebula is probably over two light-years wide, appearing quite faint, but just visible in a 102-mm (4-inch) telescope. Use low magnification to spot this one.

Mira (Omicron (o) Ceti)

TYPE: Variable Star

APPARENT MAGNITUDE: +2.00 to +10.10

POSITION: RA: 02h 19m 21s,
Dec: -02° 58' 36"

DISTANCE: 300 light-years

DESCRIPTION: Mira, a bright red giant, is losing its mass, some of which is being gravitationally pulled around its white dwarf companion. As it pulsates, its brightness varies over the course of a year or so. At this stage of its lifecycle, it has exhausted its supply of core hydrogen and is now fusing helium to form carbon and oxygen.

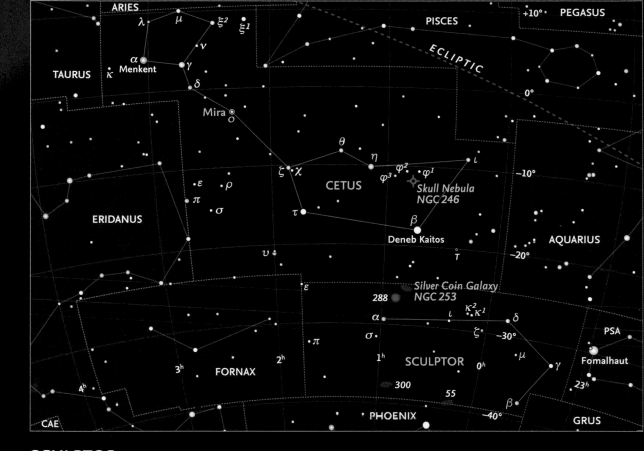

SCULPTOR

Silver Coin Galaxy (NGC 253)

TYPE: *Spiral Starburst Galaxy*

APPARENT MAGNITUDE: *+8.00*

ANGULAR SIZE: *27.5 arcmin*

POSITION: *RA: 00h 47m 33s,*
Dec: -25° 17' 18"

DISTANCE: *11.4 million light-years*

DESCRIPTION: The Silver Coin is angled almost edge-on relative to our line of sight and is a great target for small telescopes. It is undergoing intense star formation in its spiral arms possibly triggered by a collision with a dwarf galaxy some 200 million years ago. The impact injected huge amounts of gas into the galaxy, kickstarting a stellar 'baby boom'.

AQUARIUS, AQUILA AND DELPHINUS

Aquarius (Water Bearer) contains many interesting stars, but in the ancient world was looked upon unfavourably due to its perceived connection to floods. Aquila (Eagle) is home to Altair (Alpha Aquilae), the star at the southern tip of the Summer Triangle. The lovely little constellation Delphinus (Dolphin) is said to have saved the great Greek poet Arion after he was kidnapped by pirates.

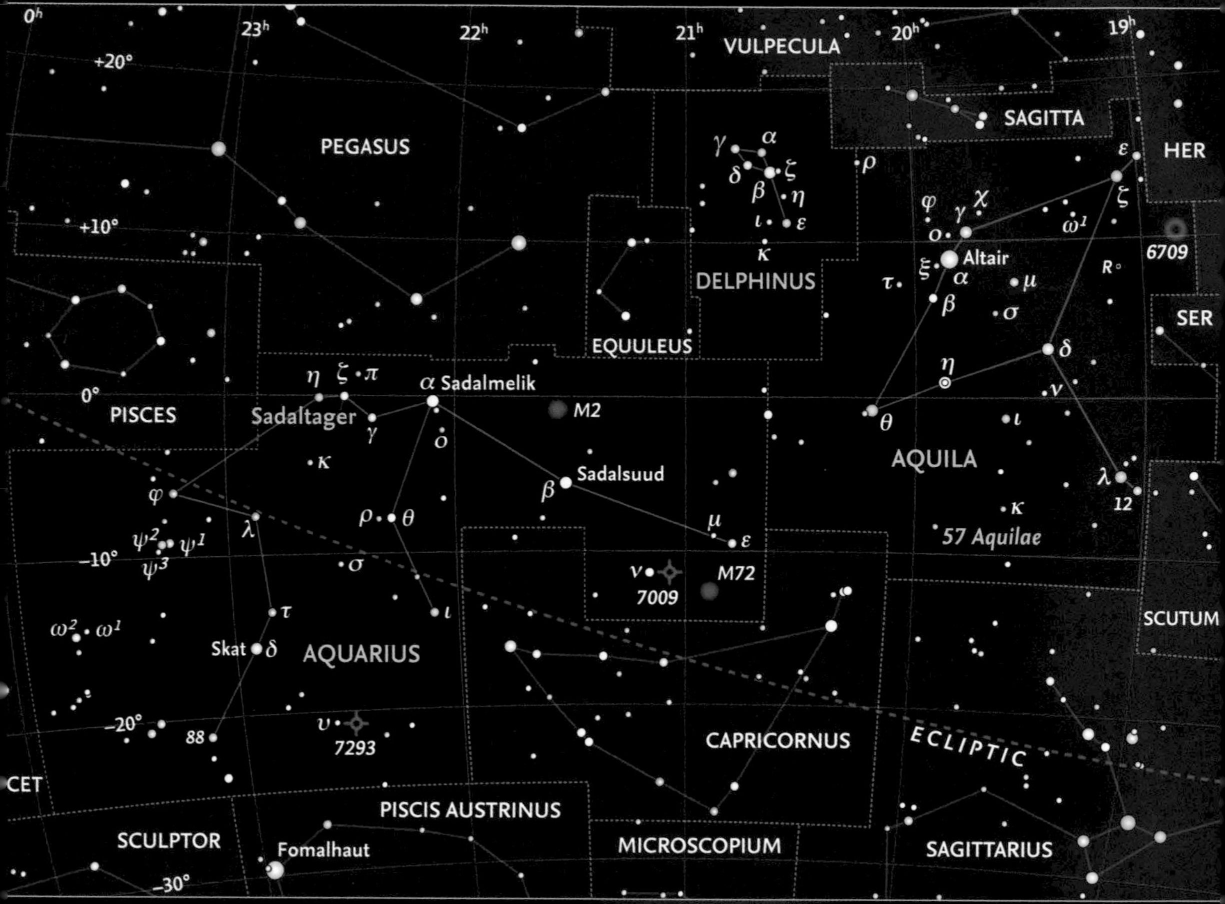

AQUARIUS

Sadaltager (Zeta (ζ) Aquarii)

TYPE: *Double Star*

APPARENT MAGNITUDE: *+4.42 / +4.51*

POSITION: *RA: 22h 28m 50s,*
Dec: -00° 01' 12"

DISTANCE: *92 light-years*

DESCRIPTION: This pleasing double star has two components of very similar brightness and colour separated by just a few seconds of arc. It's a beautiful sight at high power, but requires a telescope to see both stars separately.

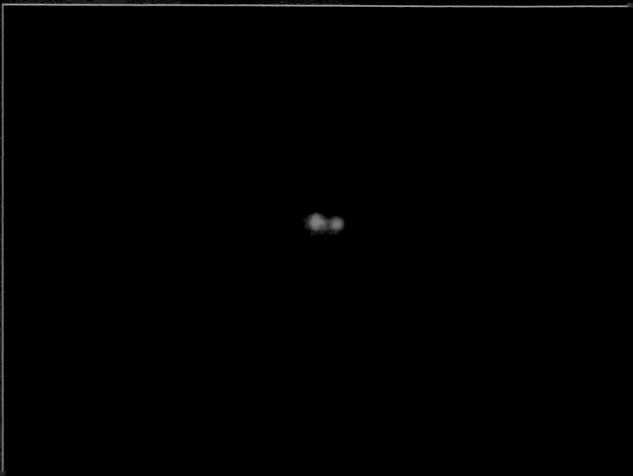

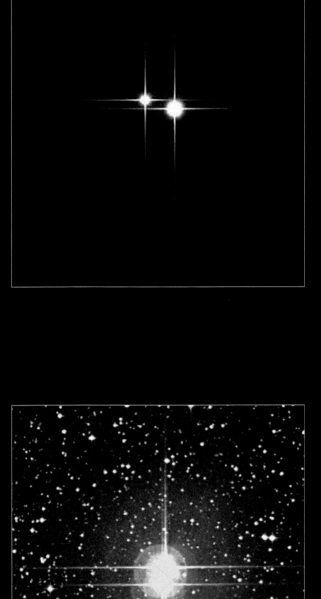

Gamma (γ) Delphini

TYPE: *Double Star*

APPARENT MAGNITUDE: *+5.14 / +4.27*

POSITION: *RA: 20h 46m 39s,*
Dec: +16° 07' 38"

DISTANCE: *101 light-years*

DESCRIPTION: The primary star of this binary is a yellow-white star orbiting a brighter orange star. The primary star is more massive than the secondary, despite being dimmer, and both stars are at least one and a half times more massive than the Sun.

AQUILA

57 Aquilae

TYPE: *Double Star*

APPARENT MAGNITUDE: *+5.70*

POSITION: *RA: 19h 54m 38s,*
Dec: -08° 13' 38.2"

DISTANCE: *500 light-years*

DESCRIPTION: The two massive B-type stars of 57 Aquilae are wide apart but may form a binary, resulting in a beautiful pair of stars. On average, the pair are about 500 light-years away, and are separated in brightness by less than one magnitude. 57 Aquilae can be split easily at relatively low telescopic powers, or even with large binoculars.

Sagittarius (Archer) and Capricornus (Sea Goat) are both zodiacal constellations. Sagittarius contains the asterism known as the Teapot, which points the way to the exact centre of our Milky Way. Capricornus is quite faint, but nevertheless easy to trace out in the sky.

SAGITTARIUS

Lagoon Nebula (M8)

TYPE: *Emission Nebula*

APPARENT MAGNITUDE: *+6.00*

ANGULAR SIZE: *90 arcmin*

POSITION: *RA: 18h 03m 37s,*
               Dec: -24° 23' 12"

DISTANCE: *4,100 light-years*

DESCRIPTION: M8 is a giant interstellar cloud of gas (mostly hydrogen) and dust, which is collapsing in many regions to form young solar systems. Its enormous billowing clouds are so bright that it is faintly visible to the naked eye. Binoculars reveal its true shape, and telescopes can tease out various structures. It's over 100 light-years wide.

M22

TYPE: *Globular Cluster*

APPARENT MAGNITUDE: *+5.10*

ANGULAR SIZE: *32 arcmin*

POSITION: *RA: 18h 36m 24s,*
               Dec: -23° 54' 17"

DISTANCE: *10,600 light-years*

DESCRIPTION: M22 is a particularly bright cluster, although from the northern hemisphere it never rises high in the sky, so its light suffers from extensive atmospheric distortion. It is one of four globular clusters known to harbour a planetary nebula: the one in M22 is called GJJC1 and it is relatively young. Its star evolved from the red giant phase and started losing mass only 6,000 years ago.

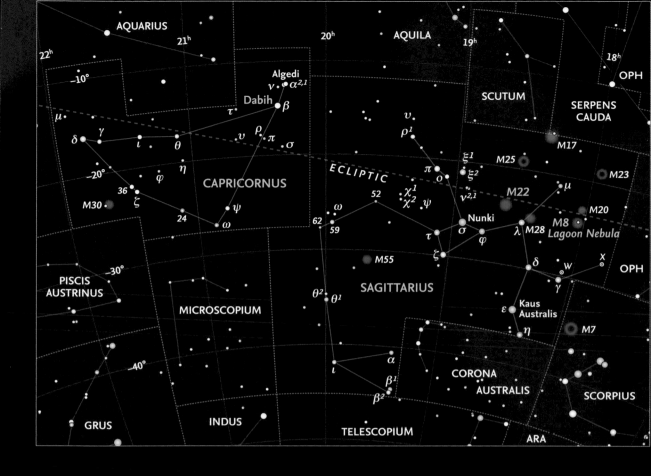

CAPRICORNUS

Dabih (Beta (β) Capricorni)

TYPE: *Multiple Star System*

APPARENT MAGNITUDE: *+3.05 / +6.09*

POSITION: *RA: 20h 21m 01s,*
Dec: -14° 46' 53"

DISTANCE: *328 light-years*

DESCRIPTION: Dabih looks like two stars through a small telescope or binoculars, separated by 3.5 arcminutes, equivalent to an actual distance of 0.34 light-years. They are in a very wide orbit, taking 700,000 years to complete one lap. Each of the stars has its own very close companion, but these are much harder to resolve. The brightest of the system is an orange giant, cooler than the other stars but much bigger and more luminous.

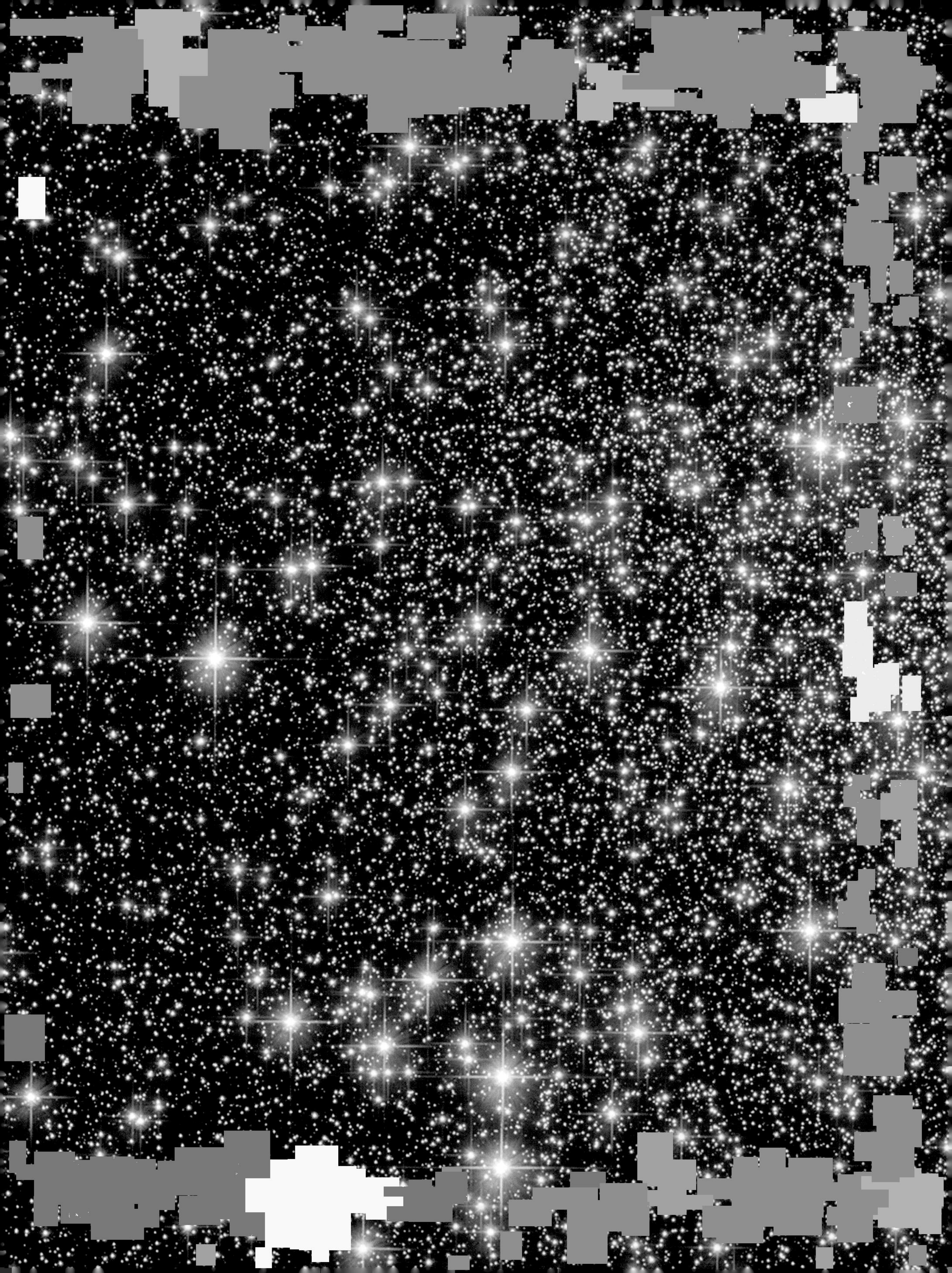

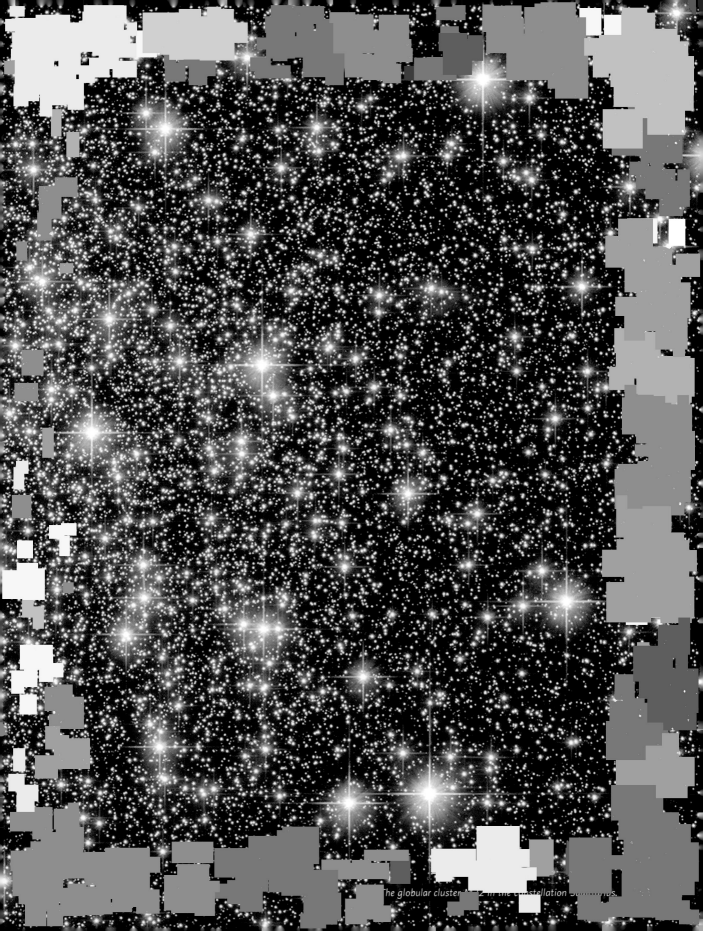
the globular cluster M 72 in the constellation Sagittarius.

OPHIUCHUS, SERPENS AND SCUTUM

Illustrations of Ophiuchus (Serpent Bearer) usually show a figure holding Serpens (Serpent). Serpens itself is divided into two regions: Serpens Cauda (the tail) and Serpens Caput (the head). Ophiuchus is sometimes called the thirteenth zodiacal constellation, because the Sun passes through it in December. Scutum (Shield) is a faint constellation with no stars brighter than the third magnitude.

SERPENS

M5

TYPE: *Globular Cluster*

APPARENT MAGNITUDE: *+6.65*

ANGULAR SIZE: *23 arcmin*

POSITION: *RA: 15h 18m 33s,*
Dec: +02° 04' 51.7"

DISTANCE: *24,500 light-years*

DESCRIPTION: M5 is a large globular cluster containing up to half a million stars densely packed into a sphere 165 light-years in diameter. It is full of old stars: the cluster formed only 800 million years after the birth of the Universe.

OPHIUCHUS

61 Ophiuchi

TYPE: *Double Star*

APPARENT MAGNITUDE: *+6.2*

POSITION: *RA: 17h 44m 34s,*
Dec: +02° 34' 46"

DISTANCE: *460 light-years*

DESCRIPTION: This pair of stars is separated by about 20 arcseconds, making it an easy split for small telescopes at medium to high magnifications. The stars are both of similar spectral type (A) and apparent brightness. Although at first glance the stars appear white, many observers agree they exhibit a very pale yellow tint, which can be glimpsed with careful study. Can you see it?

SCUTUM

Wild Duck Cluster (M11)

TYPE: *Open Cluster*

APPARENT MAGNITUDE: *+6.30*

ANGULAR SIZE: *14 arcmin*

POSITION: *RA: 18h 51.1m,*
Dec: -06° 16'

DISTANCE: *6,000 light-years*

DESCRIPTION: Several thousand stars populate this rich, open cluster, which lies roughly 6,000 light-years away. Its overall density, combined with its distance, gives it a compact, busy appearance. M11 is easily visible in binoculars, so it's no surprise it was discovered all the way back in the seventeenth century. The cluster's age has been estimated to be anywhere between 220 million to 500 million years old.

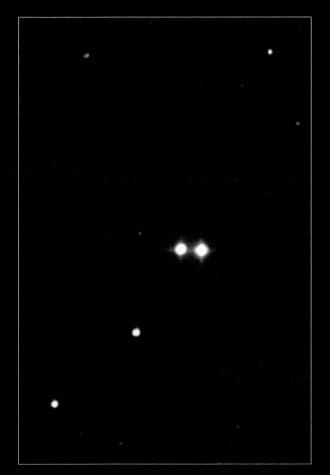

CENTAURUS AND SCORPIUS

Centaurus (Centaur) contains the nearest-known star to the Sun, Rigel Kentaurus (Alpha Centauri), which is actually a system of several stars about four light-years away. The nearby Scorpius (Scorpion) is a member of the Zodiac and is a spectacular constellation from its head to its sting. It's also home to Antares, a bright red star known as the rival of Mars.

SCORPIUS

M4

TYPE: *Globular Cluster*

APPARENT MAGNITUDE: *+5.90*

ANGULAR SIZE: *26 arcmin*

POSITION: *RA: 16h 23m 35s,*
Dec: -26° 31' 37"

DISTANCE: *7,200 light-years*

DESCRIPTION: Right next to Antares, the heart of the scorpion, we find M4: a globular cluster comprising hundreds of thousands of stars. It's around 75 light-years wide, and at 7,000 light-years from us, it is one of the nearer globular clusters. It's estimated to be over 12 billion years old.

Ptolemy's Cluster (M7)

TYPE: *Open Cluster*

APPARENT MAGNITUDE: *+3.30*

ANGULAR SIZE: *80 arcmin*

POSITION: *RA: 17h 53m 51s,*
Dec: -34° 47' 34"

DISTANCE: *980 light-years*

DESCRIPTION: Visible to the naked eye, this cluster harbours almost 100 stars and is about 200 million years old. It covers about 1.3 degrees in the sky, and its actual diameter is 25 light-years. It's the southernmost object in Charles Messier's famous catalogue of objects.

CENTAURUS

Omega (ω) Centauri (NGC 5139)

TYPE: *Globular Cluster*

APPARENT MAGNITUDE: *+3.90*

ANGULAR SIZE: *36.3 arcmin*

POSITION: *RA: 13h 26m 47s,*
Dec: -47° 28' 46"

DISTANCE: *15,100 light-years*

DESCRIPTION: In 1677, Edmund Halley discovered what astronomers now know to be the most massive globular cluster in the Milky Way's neighbourhood. Naturally, it is considered by many to be the most spectacular in the sky, and is actually quite visible to the naked eye. Thought to be a core remnant of a failed dwarf galaxy, it is home to millions of stars almost twice the age of the Sun, over 11 billion years! At about 100x magnification in a telescope, it fills your view like a fountain of stars. Omega Centauri is a true highlight of the southern skies.

HYDRA AND ANTLIA

Hydra (Water Snake) is the largest constellation in the entire sky, but home to just one first magnitude star, Alphard (Alpha Hydrae). One of its neighbouring constellations is the faint Antlia (Air Pump).

HYDRA

M48

TYPE: *Open cluster*

APPARENT MAGNITUDE: *+5.50*

ANGULAR SIZE: *54 arcmin*

POSITION: *RA: 08h 13.7m,*
Dec: -05° 45'

DISTANCE: *1,500 light-years*

DESCRIPTION: The magnitude of this cluster is close to the visual limit of the eye, but it can be seen in very dark clear skies using averted vision.

Southern Pinwheel Galaxy (M83)

TYPE: *Spiral Galaxy*

APPARENT MAGNITUDE: *+7.54*

ANGULAR SIZE: *12.9 arcmin*

POSITION: *RA: 13h 37m 01s,*
Dec: -29° 51' 57"

DISTANCE: *15 million light-years*

DESCRIPTION: 15 million light-years away, this barred spiral galaxy shines with the light of hundreds of billions of suns. Looking face on, we see a galaxy like our own Milky Way, with its pronounced spiral arms visible in larger instruments. M83 is one of the brightest galaxies in the summer skies.

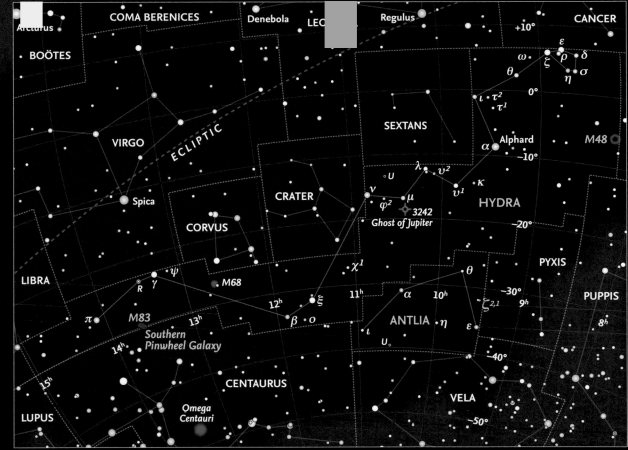

The star map shows the following labels and constellations:

COMA BERENICES · Denebola · LEO · Regulus · CANCER
Arcturus · BOÖTES
+10°
ε
ω · ζ ρ · δ
θ · η σ
0°
ι · τ²
· τ¹
SEXTANS
VIRGO
ECLIPTIC
α · Alphard · M48
−10°
λ · υ²
∘ U
ν · κ
CRATER · υ¹
Spica · HYDRA
μ
φ² 3242
Ghost of Jupiter
−20°
CORVUS
χ¹
θ · PYXIS
LIBRA · M68
11ʰ · α · 10ʰ · −30°
R · γ · ψ · ζ²,¹ · 9ʰ · PUPPIS
12ʰ · ξ
π · M83 · β · ο · ANTLIA · η
13ʰ · ι · ε · 8ʰ
Southern · U∘
Pinwheel Galaxy · −40°
14ʰ
15ʰ · CENTAURUS · VELA
LUPUS · Omega · −50°
Centauri

ANTLIA

Zeta (ζ) Antliae

TYPE: *Double Star*

APPARENT MAGNITUDE: *+5.76*

ANGULAR SIZE: *54 arcmin*

POSITION: *RA: 09h 30m 46s,
Dec: -31° 53' 22"*

DISTANCE: *410 light-years*

DESCRIPTION: Both stars are yellow-white and hotter than the Sun, and are 8 arcseconds apart, easily seen through a telescope of any size. Though subtly different in brightness, the two stars are virtually identical in colour.

VELA AND PUPPIS

Once part of the great ship Argo Navis, Vela (Sails) and Puppis (Poop Deck) are rich constellations that can be seen in front of the Milky Way. Scanning this area of the sky with binoculars is very rewarding, as you'll find many clusters and pools of light.

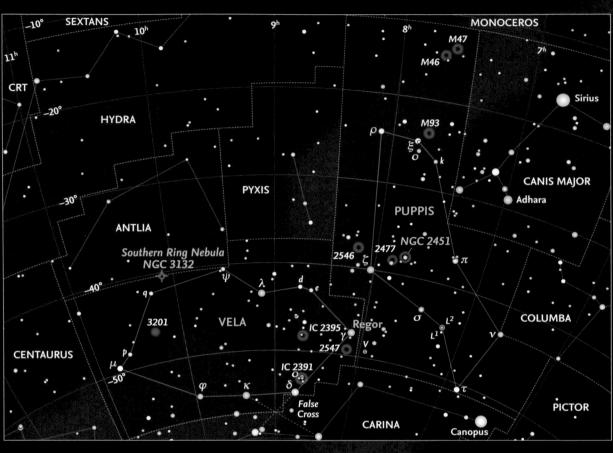

VELA

Southern Ring Nebula (NGC 3132)

TYPE: *Planetary Nebula*

APPARENT MAGNITUDE: *+9.87*

ANGULAR SIZE: *62 arcmin*

POSITION: *RA: 10h 07m 02s,
Dec: -40° 26' 11"*

DISTANCE: *2,000 light-years*

DESCRIPTION: The Southern Ring Nebula is a spectacular nebula around a hot white dwarf star, whose radiation excites and heats the surrounding gas that once made up the outer shell of a star similar to the Sun. The nebula is nearly half a light-year across.

Regor (Gamma (γ) Velorum)

TYPE: Multiple Star System

APPARENT MAGNITUDE: +1.83

POSITION: RA: 08h 09m 32s,
Dec: -47° 20' 12"

DISTANCE: 1,115 light-years

DESCRIPTION: Regor is an incredibly bright star system composed of a Wolf-Rayet star and three companions, one of which is a blue supergiant 30 times more massive than the Sun. Wolf-Rayet stars are at least five times hotter than the Sun and produce intense stellar winds – they lose their outer layers of hydrogen and helium as they evolve, often transferring this mass to a stellar companion. The Wolf-Rayet star in this system is nine solar masses and will eventually explode as a supernova.

PUPPIS

NGC 2451

TYPE: Open Cluster

APPARENT MAGNITUDE: +3.00

ANGULAR SIZE: 45 arcmin

POSITION: RA: 07h 45m 24s,
Dec: -37° 57' 00"

DISTANCE: 600 light-years, 1,200 light-years

DESCRIPTION: This lovely open cluster in Puppis has been known for centuries, but only recently it was found to be more complicated than it seemed. NGC 2451 is in fact a pair of clusters (known as A and B) along the same line of sight. The nearer cluster is 600 light-years away, whereas its neighbour is twice as far at 1,200 light-years. It's easily seen with binoculars – look out for a bright, distinctly orange star.

Carina (Keel) was the largest section of the ancient Argo Navis constellation. It's home to the very bright star Canopus (Alpha Carinae) and many deep sky objects. You can easily spend many nights exploring this splendid area of the sky.

Eta Carinae

TYPE: *Multiple Star System,*
Emission Nebula

APPARENT MAGNITUDE: *-1.00 to +7.60*

POSITION: *RA: 10h 45m 04s,*
Dec: -59° 41' 04"

DISTANCE: *7,500 light-years*

DESCRIPTION: This star system brightened in an event called the Great Eruption in the mid-1800s. Its two known stars orbit each other every 5.5 years. One is a luminous blue variable, which was up to 250 solar masses before it underwent significant mass loss through its strong stellar winds. The other is a blue supergiant up to 80 times the mass of the Sun. The system is enshrouded in gas and dust from the Great Eruption, making it difficult for astronomers to study the stars within. Eta Carinae is embedded within a very interesting region of the sky: the surrounding nebulae and star clusters are a magnificent sight and well worth exploring.

Southern Pleiades (IC 2602)

TYPE: *Open Cluster*

APPARENT MAGNITUDE: *+1.90*

ANGULAR SIZE: *50 arcmin*

POSITION: *RA: 10h 42m,*
Dec: -64° 23'

DISTANCE: *460 light-years*

DESCRIPTION: This cluster is also known as the Theta Carinae Cluster, and although it does not quite match the Pleiades in splendour, it does put up a good fight. Its distance from us is similar to that of the well-known Taurean cluster and is composed of bright, hot blue stars a few tens of millions of years old. It's a grand sight in binoculars, and can be seen quite easily with the naked eye.

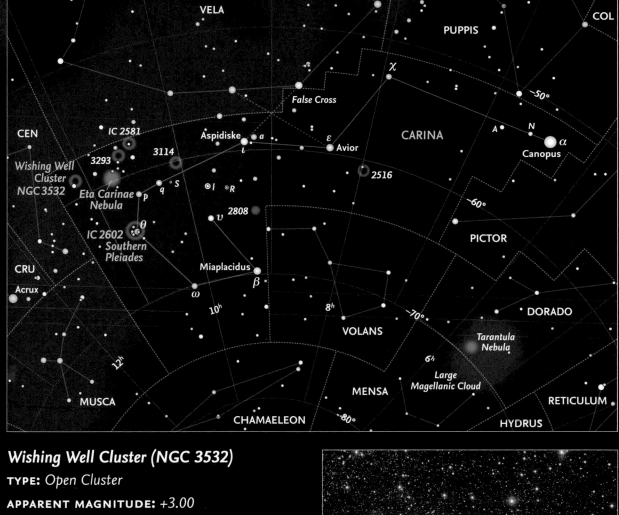

Wishing Well Cluster (NGC 3532)

TYPE: *Open Cluster*

APPARENT MAGNITUDE: +3.00

ANGULAR SIZE: 55 arcmin

POSITION: *RA: 11h 05m 33s,*
Dec: -58° 44' 1"

DISTANCE: *1,321 light-years*

DESCRIPTION: We agree with the nineteenth-century astronomer John Herschel's claim that this is one of the finest star clusters in the whole sky. It has the honour of being the first object observed with the Hubble Space Telescope, and you can track it down for yourself. Its 150 evenly spread members cover almost a whole degree of sky, so high magnifications are not required to admire this object. However, with enough power it's possible to get completely lost among its stars, though that's no bad thing!

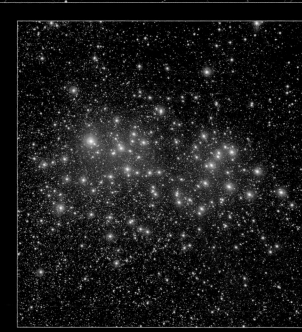

The Eta Carinae region in the constel_____ _arina. **165**

Pavo (Peacock), Telescopium (Telescope) and Ara (Oven) are all quite faint, although Pavo does contain one first magnitude star, also named Peacock (Alpha Pavonis). Of the three, only Ara was featured in Ptolemy's ancient table of 48 constellations, and the others were created a few centuries ago.

PAVO

NGC 6752

TYPE: *Globular Cluster*

APPARENT MAGNITUDE: *+5.40*

ANGULAR SIZE: *20.4 arcmin*

POSITION: *RA: 19h 10m 52s,*
Dec: -59° 59' 04.4"

DISTANCE: *13,000 light-years*

DESCRIPTION: NGC 6752 is the third brightest globular cluster in the night sky, with a densely packed core of crowded stars. Due to the close proximity of the stars, there are many binary systems and blue stragglers – these large bright stars may be the product of two stars colliding and merging to form a giant star.

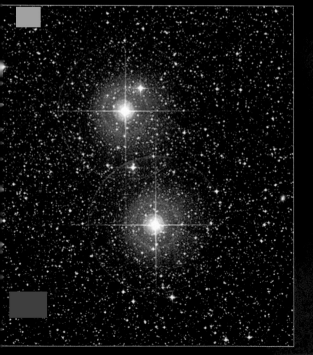

TELESCOPIUM

Delta (δ) Telescopii

TYPE: *Line of Sight Double Star*

APPARENT MAGNITUDE: *+4.90 / +5.05*

POSITION: *RA: 18h 31m 45s,*
Dec: -45° 54' 54"
RA: 18h 32m 02s,
Dec: -45° 45' 26"

DISTANCE: *710 light-years, 1,200 light-years*

DESCRIPTION: These two stars have a wide separation of 0.16 degrees (9.6 arcminutes). They look like a binary, but in fact they are situated at significantly different distances from the Sun – they are an optical double. One of the stars is a massive blue-white supergiant.

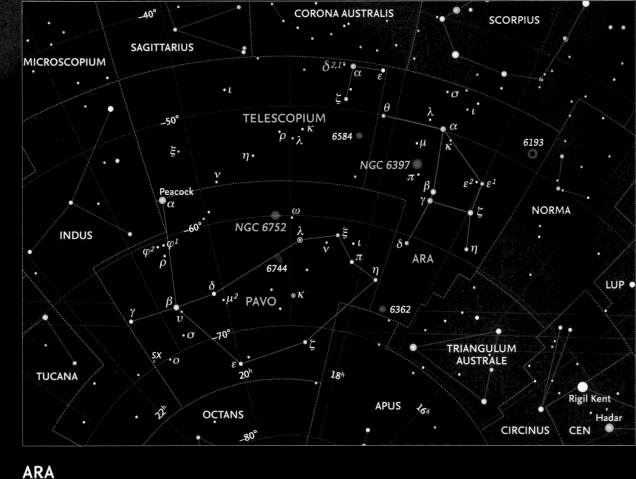

ARA

NGC 6397

TYPE: *Globular Cluster*

APPARENT MAGNITUDE: *+6.68*

ANGULAR SIZE: *32 arcmin*

POSITION: *RA: 17h 40m 42s,*
Dec: -53° 40' 28"

DISTANCE: *7,200 light-years*

DESCRIPTION: This huge cluster of 400,000 stars has undergone core collapse, whereby some stars have migrated outwards and the remaining stars in the centre have collapsed into a more compact volume causing overcrowding. Close binary systems form in this region. The cluster itself is 13.6 billion years old, and formed only 200 million years after the Big Bang.

Crux (Cross) is the smallest of all the 88 modern constellations. It's a very recognisable sight in the southern skies, home to the famous Jewel Box Cluster. Volans (Flying Fish) is one of several somewhat faint southern constellations.

CRUX

Jewel Box Cluster (NGC 4755)

TYPE: *Open Cluster*

APPARENT MAGNITUDE: *+3.00*

ANGULAR SIZE: *55 arcmin*

POSITION: *RA: 12h 53m 42s,*
Dec: -60° 22'

DISTANCE: *6,400 light-years*

DESCRIPTION: John Herschel described this cluster as 'a superb piece of fancy jewellery', giving it a famous name and rendering it a firm favourite in the southern skies. It is only faintly visible to the naked eye, and needs magnification to draw out its 100 or so young stars. Herschel remarked upon the striking range of colours visible among these stars.

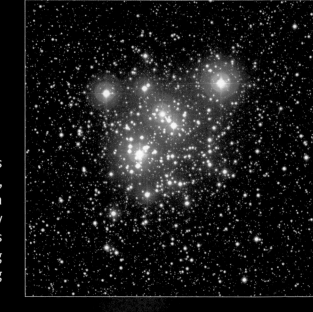

Coalsack Nebula

TYPE: *Dark Nebula*

ANGULAR SIZE: *7 deg*

POSITION: *RA: 12h 50m,*
Dec: -62° 30'

DISTANCE: *600 light-years*

DESCRIPTION: Not all nebulae are illuminated by starlight. Dark nebulae such as the Coalsack are made up of interstellar dust so dense that they block out the light behind them, and appear like silhouettes or holes in the galaxy. This is one of the best examples of a dark nebula, and it is easily seen with the naked eye. With binoculars, it has an eerily empty appearance. Perhaps one day, regions of this nebula will light up as stars begin to form inside it.

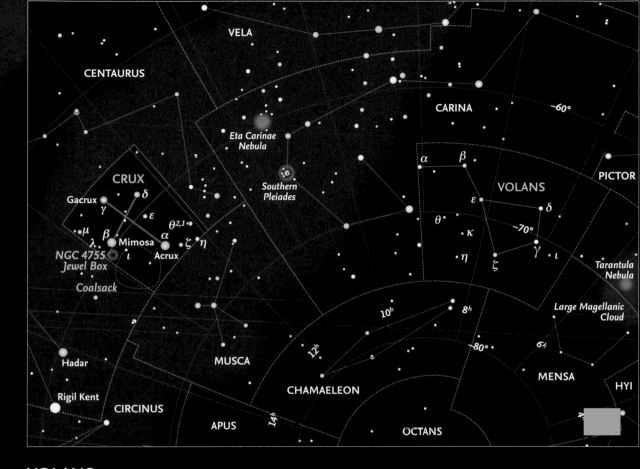

VOLANS

Gamma (γ) Volantis

TYPE: *Double Star*

APPARENT MAGNITUDE: *+3.78 / +5.68*

POSITION: *RA: 07h 08m 45s,*
Dec: -70° 29' 56"

DISTANCE: *142 light-years*

DESCRIPTION: The primary star of this binary is an orange giant, larger and cooler than the Sun. It is locked in an orbital dance with a yellow-white star 14.1 arcseconds away in the sky, and the pair can be resolved with a small telescope. Spotting the subtle colour difference between the two stars is a rewarding challenge!

TUCANA AND DORADO

Tucana (Toucan) and Dorado (Dolphinfish) are notable for containing the largest portions of the Small and Large Magellanic Clouds – satellite galaxies of the Milky Way which are visible to the naked eye.

TUCANA

Small Magellanic Cloud

TYPE: *Irregular Dwarf Galaxy*

APPARENT MAGNITUDE: *+2.70*

ANGULAR SIZE: *5.33 deg*

POSITION: *RA: 00h 52m 45s,*
Dec: -72° 49' 43"

DISTANCE: *197,000 light-years*

DESCRIPTION: The Small Magellanic Cloud (SMC) can be seen clearly with the naked eye from the southern hemisphere. It is an irregular dwarf galaxy containing a few hundred million stars. It is gravitationally bound to the Milky Way and to the Large Magellanic Cloud. A bridge of gas between the two suggests a tidal interaction: their mutual gravitational pull funnels gas across and triggers star formation within this bridge. The bright young open cluster NGC 346 lies within the heart of the SMC.

47 Tucanae

TYPE: *Globular Cluster*

APPARENT MAGNITUDE: *+4.91*

ANGULAR SIZE: *30.9 arcmin*

POSITION: *RA: 00h 52m 45s,*
Dec: -72° 49' 43"

DISTANCE: *16,700 light-years*

DESCRIPTION: Southern observers are spoiled for choice when it comes to spectacular globular clusters. 47 Tucanae is bright enough to be visible to the naked eye, and close enough to occupy about half a degree of sky. This cluster has provided some scientific surprises, such as the range of exotic stars in its core, and its apparent lack of exoplanets (planets orbiting other stars), despite expectations that it would contain at least a few systems.

DORADO

Large Magellanic Cloud

TYPE: *Irregular Dwarf Galaxy*

APPARENT MAGNITUDE: *+0.90*

ANGULAR SIZE: *10.75 deg*

POSITION: *RA: 05h 23m 34.5s,
Dec: -69° 45' 22"*

DISTANCE: *163,000 light-years*

DESCRIPTION: The Large Magellanic Cloud (LMC) is a satellite galaxy of the Milky Way, gravitationally bound to it along with the Small Magellanic Cloud. It has a mass of 10 billion Suns and appears to have a bright bar running through its core and a spiral arm. However, because it has no prominent structure, it is classified as an irregular galaxy – its spiral nature may have been disrupted by its interaction with the Milky Way. Active star formation occurs in the LMC, particularly in the Tarantula Nebula. It is thought that if this nebula was as close to us as the Orion Nebula, its intense light would cast shadows on the surface of the Earth. The LMC also houses at least 60 globular clusters, as well as stars at all stages of their evolutionary lifecycles.

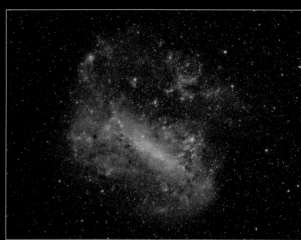

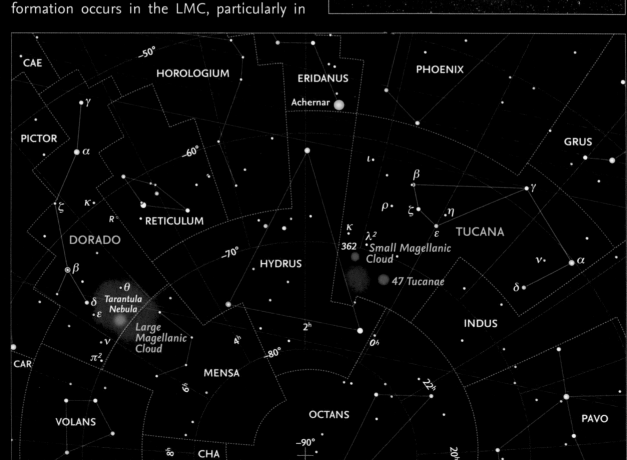

START STARGAZING!

Aurora Borealis.   173

Orion with its prominent belt and nebula as seen from Cerro Paranal in Chile.

TRIP ONE

Constellations and Stars

Now that you are ready to enjoy the night sky, follow our tips for planning your first three stargazing sessions – just enough to get you hooked for life!

Start with just your eyes on your first trip – it's important you learn to find your way around the sky before bringing binoculars and a telescope. During your first trip, you'll be looking for constellations and asterisms and using them to find your direction. Follow the checklist below to plan your trip:

- Choose a suitable location (ideally with a clear view of the horizon, the darker the better).

- Check the weather forecast.

- Check sunset and moonrise/moonset times. To see the stars, the Moon should be below the horizon.

- Use the constellation charts in this book and Stellarium or another app to see which constellations are visible from your location after sunset. The higher the altitude of the stars, the sharper they'll be.

- Make copies of the observing log in this book and take them with you, along with a pencil and clipboard.

- Take a red torch if you are heading somewhere dark, and wear warm clothes.

- Once you are at your location allow, at least 30 minutes for your eyes to adapt to the dark. If you are using an app on your phone, switch on the night mode function.

- Use the circumpolar pointer charts in this book to help you find the north or south celestial pole. Look directly down from the celestial poles, so that you are facing north or south. Try finding other circumpolar constellations in this part of the sky, and sketch what you see.

- Turn around and face the opposite direction. Sketch the positions of the stars in the brightest constellations. If you are in a dark region outside of the city, challenge yourself by looking for fainter stars, and count how many you see in each constellation. Try looking for the Milky Way in darker clear skies.

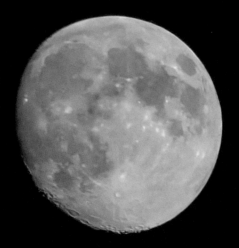

TRIP TWO

Moon and Planets

Now you know your way around the sky, you can look at the Moon and planets and sketch what you see in your observing log (or better still, take some photos). You may wish to take binoculars and a tripod with you – this will allow you to see the craters on the Moon and to see the moons around Jupiter or Saturn. Follow the checklist below when planning your trip:

- Choose a suitable location with a clear view of the sky opposite to the celestial pole (face south if you live in the northern hemisphere, and face north if you live in the southern hemisphere).

- Check the weather forecast.

- Check sunset and moonrise/moonset times. The best time to look at the Moon through a telescope is in its crescent, quarter or gibbous phase. You will see shadows cast by the crater walls near the terminator.

- Use Stellarium or another app to see which planets are visible that night and when the Moon and planets will be on the meridian.

- Make copies of the observing log in this book and take them with you, along with a pencil and clipboard.

- Take a red torch if you are heading somewhere dark, and wear warm clothes.

- Set up your binoculars or camera on a tripod. Make sure your binoculars are focused by turning the diopter ring (in the middle) – try looking at the Moon first and some of its big craters like Kepler, Tycho and Copernicus. This will help you with focusing. You may want to find the Apennine mountain range by the Mare Imbrium (Sea of Showers).

- Use a star app on your phone or printed star charts from Stellarium to locate planets – knowing which constellation they are in will help you to find them. Use your binoculars to look for the four largest moons around Jupiter. Depending on your magnification, you may even spot some of the moons around Saturn.

The globular star cluster 47 Tucanae in the constellation Tucana.

TRIP THREE

Clusters, Nebulae and Galaxies

Now it's time to look for more challenging objects like emission nebulae around star-forming clusters, planetary nebulae, open and globular clusters and other galaxies. You can try taking photos with your smartphone, but to capture more detail and colours of the nebulae you will need a DSLR camera. Follow the checklist below when planning your trip:

- Choose a suitable location (ideally with a clear view of the horizon, the darker the better).

- Check the weather forecast.

- Check sunset and moonrise/moonset times. To see deep sky objects, the Moon should be below the horizon.

- Use the constellation charts in this book and Stellarium or another app to see which constellations and objects are visible from your location after sunset.

- Make copies of the observing log in this book and take them with you, along with a pencil and clipboard.

- Take a red torch if you are heading somewhere dark, and wear warm clothes.

- Once you are at your location allow 30 minutes for your eyes to adapt to the dark. If you are using an app on your phone, switch on the night mode function.

- Now you can start looking for constellations and a few deep sky objects in each one. Start with the brighter clusters and nebulae, and aim for an apparent magnitude of 4 or brighter. You may want to use binoculars or a telescope: set these up on a tripod and try focusing your equipment on individual stars before looking for faint diffuse nebulae.

- A DSLR camera can be attached to the eyepiece of your telescope, allowing you to take steady long exposure images of your chosen object. Remember to move your telescope in order to track each object as the sky moves. You can either do this manually or electronically, depending on your telescope – telescopes with an equatorial mount can track the sky by themselves. Once you are happy using your telescope you can try looking for fainter objects – a successful observing session takes lots of practice and patience, and the results are always incredibly rewarding.

Icy structures called penitentes in the Atacama Desert, Chile overlooked by the constellation Orion, with Sirius just above the horizon and Canopus further right. **181**

TAURUS

AURIGA

Capella

Mira

CETUS

Pleiades

ECLIPTIC

PERSEUS

ARIES

Algol

PISCES

TRIANGULUM

ANDROMEDA

Double Cluster

CAMELOPARDALIS

M31
Andromeda Galaxy

CASSIOPEIA

PEGASUS

Polar

CEPHEUS

LACERTA

Deneb

WINTER SKY

December 01	01:00
December 15	24:00
January 01	23:00
January 15	22:00

CYGNUS

LYRA

Vega

Stellar magnitudes:

M35

GEMINI

Pollux

Castor

M44
Praesepe

CANCER

Alphard

HYDRA

ZENITH

SEXTANS

LYNX

Regulus

LEO

LEO MINOR

URSA MAJOR

The Pointers

URSA MINOR

Ursids

*The Plough
(Big Dipper)*

Denebola

VIRGO

Melotte 111

CANES VENATICI

COMA
BERENICES

DRACO

M3

Quadrantids

BOÖTES

London – Looking North

February 01	21:00
February 15	20:00
March 01	19:00
March 15	18:00

HERCULES

N

-1 0 1 2 3 4 5

ECLIPTIC

E

WINTER SKY

December 01   01:00
December 15   24:00
January 01   23:00
January 15   22:00

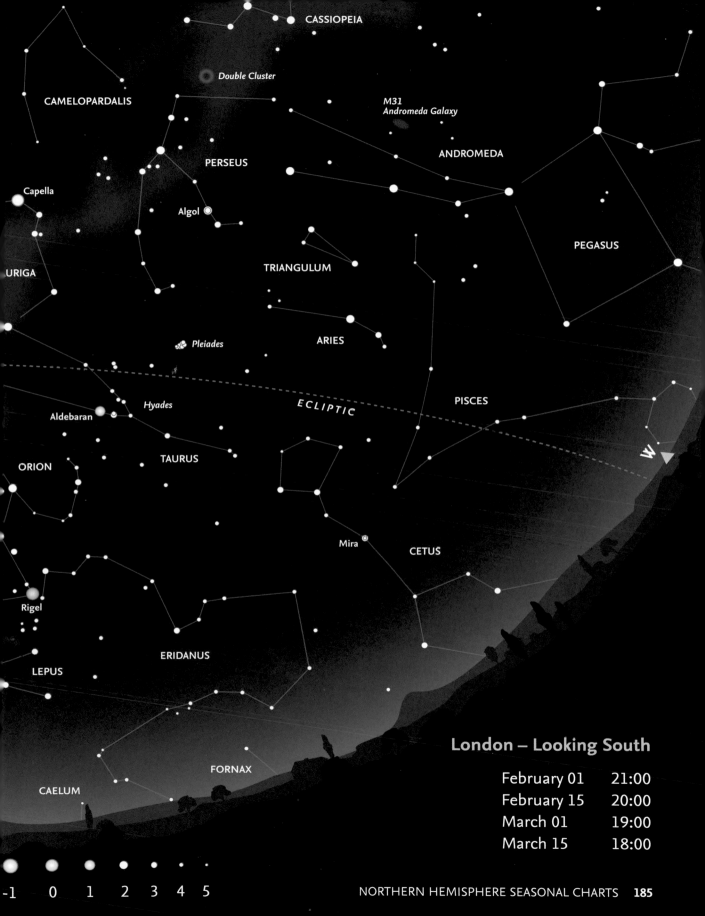

CASSIOPEIA

Double Cluster

CAMELOPARDALIS

M31
Andromeda Galaxy

ANDROMEDA

PERSEUS

Capella

Algol ◎

PEGASUS

TRIANGULUM

URIGA

ARIES

Pleiades

PISCES

Hyades

Aldebaran

ECLIPTIC

ORION

TAURUS

W

Mira ◉

CETUS

Rigel

ERIDANUS

LEPUS

London – Looking South

February 01	21:00
February 15	20:00
March 01	19:00
March 15	18:00

FORNAX

CAELUM

-1   0   1   2   3   4   5

HYDRA

LEO

LEO MINOR

M44 Praesepe

CANCER

ZENITH

Procyon

ECLIPTIC

URSA MAJOR

CANIS MINOR

Pollux

The
Pointers

Castor

GEMINI

LYNX

MONOCEROS

Betelgeuse

M35

CAMELOPARDALIS

ORION

AURIGA

Polar

Capella

TAURUS

Aldebaran

PERSEUS

CASSIOPEIA

Double Cluster

Pleiades

Algol

SPRING SKY

ANDROMEDA

TRIANGULUM

Andromeda Galaxy
M31

February 01	03:00
February 15	02:00
March 01	01:00
March 15	24:00

Stellar magnitudes

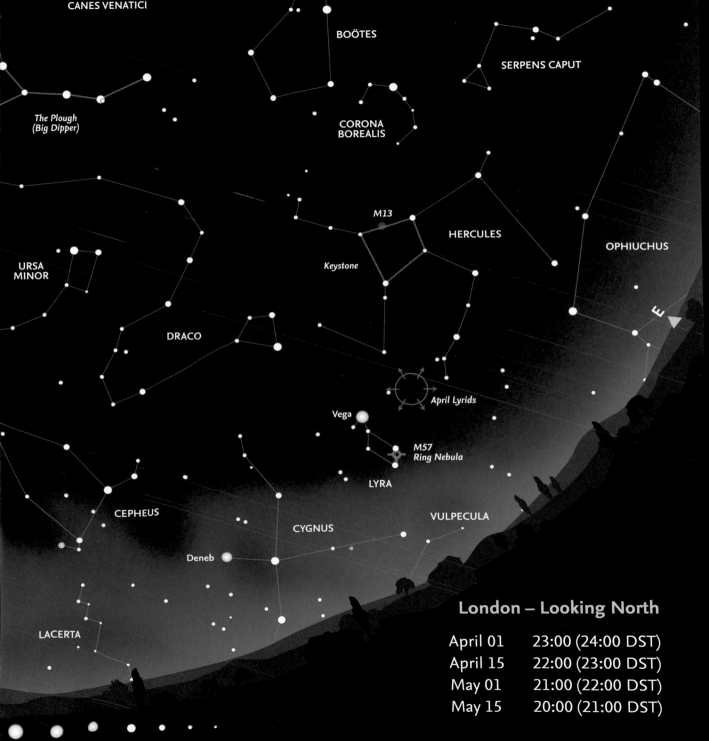

CANES VENATICI

BOÖTES

SERPENS CAPUT

*The Plough
(Big Dipper)*

CORONA
BOREALIS

M13

HERCULES

OPHIUCHUS

URSA
MINOR

Keystone

E

DRACO

April Lyrids

Vega

M57
Ring Nebula

LYRA

VULPECULA

CEPHEUS

CYGNUS

Deneb

LACERTA

London – Looking North

April 01	23:00 (24:00 DST)
April 15	22:00 (23:00 DST)
May 01	21:00 (22:00 DST)
May 15	20:00 (21:00 DST)

CYGNUS

Vega

LYRA

M57
Ring Nebula

April Lyrids

DRACO

Keystone

M13

HERCULES

CORONA
BOREALIS

BOÖTES

*The Plough
(Big Dipper)*

CANES VENATICI

M3

Melotte 111

COMA BERENICES

SERPENS
CAPUT

Arcturus

Denebola

E

OPHIUCHUS

VIRGO

Spica

CRATER

LIBRA

CORVUS

HYDRA

CENTAURUS

SPRING SKY

February 01	03:00
February 15	02:00
March 01	01:00
March 15	24:00

Stellar magni

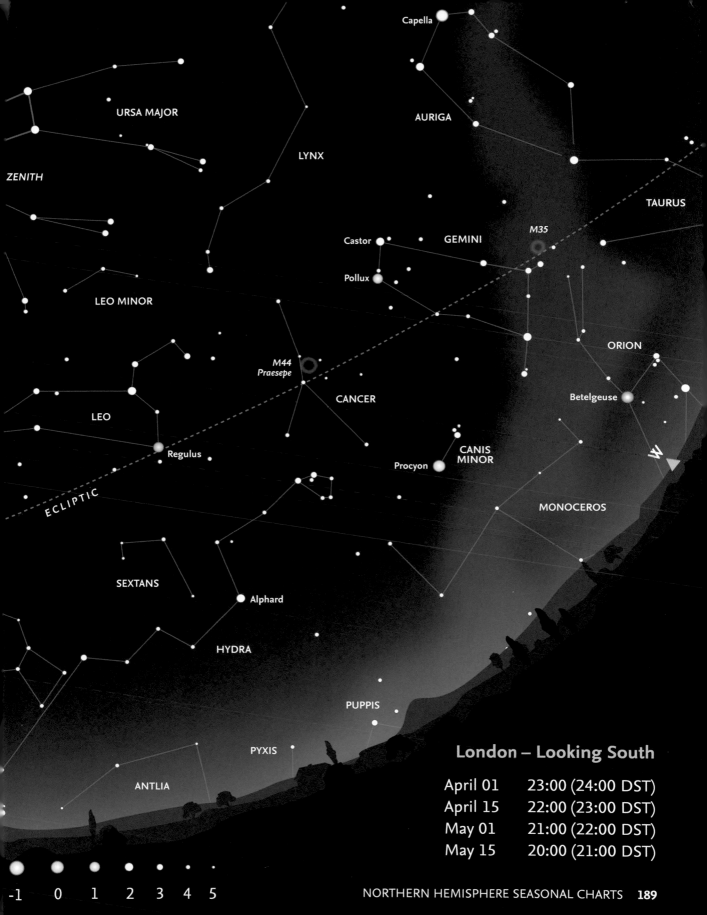

ZENITH

URSA MAJOR

LYNX

Capella

AURIGA

TAURUS

LEO MINOR

Castor

GEMINI

M35

Pollux

M44
Praesepe

CANCER

ORION

Betelgeuse

LEO

CANIS
MINOR

Regulus

Procyon

W

MONOCEROS

ECLIPTIC

SEXTANS

Alphard

HYDRA

PUPPIS

London – Looking South

PYXIS

ANTLIA

April 01	23:00 (24:00 DST)
April 15	22:00 (23:00 DST)
May 01	21:00 (22:00 DST)
May 15	20:00 (21:00 DST)

-1   0   1   2   3   4   5

SERPENS CAPUT

Keystone

CORONA
BOREALIS

M13

HERCULES

Spica

Arcturus

ZENITH

BOÖTES

ECLIPTIC

M3

VIRGO

COMA
BERENICES

DRACO

The Plough
(Big Dipper)

URSA MINOR

Melotte 111

CANES VENATICI

Denebola

URSA MAJOR

The Pointers

LEO

LEO MINOR

AURIGA

LYNX

SUMMER SKY

June 01	01:00 (02:00 DST)
June 15	24:00 (01:00 DST)
July 01	23:00 (24:00 DST)
July 15	22:00 (23:00 DST)

Stellar magnitudes:

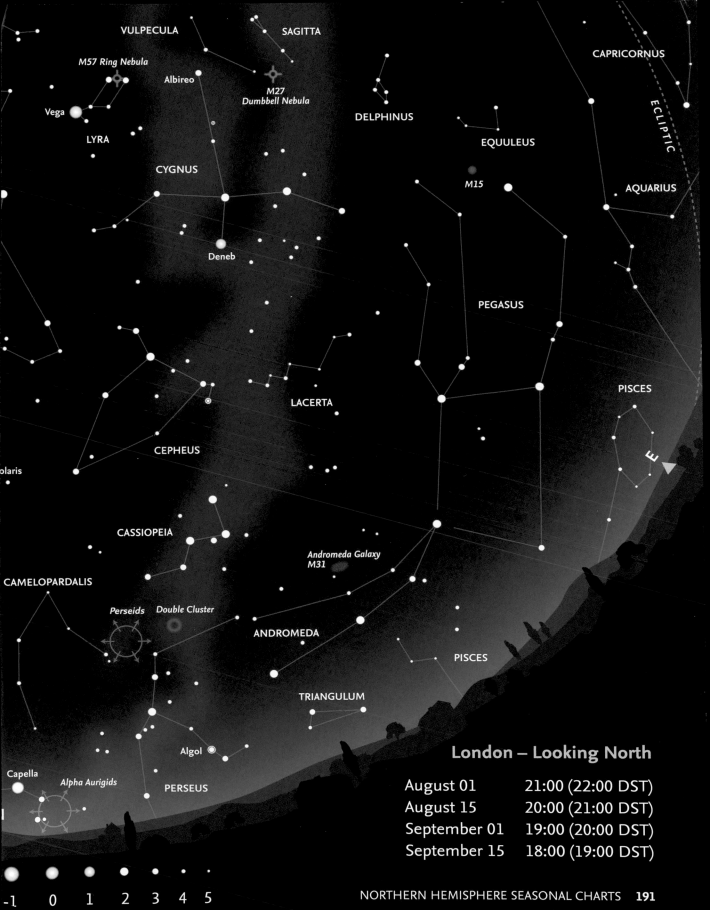

VULPECULA
SAGITTA
CAPRICORNUS
M57 Ring Nebula
Albireo
M27
Dumbbell Nebula
Vega
DELPHINUS
LYRA
EQUULEUS
AQUARIUS
CYGNUS
M15
Deneb
PEGASUS
LACERTA
PISCES
CEPHEUS
E
Polaris
CASSIOPEIA
Andromeda Galaxy
M31
CAMELOPARDALIS
Perseids   Double Cluster
ANDROMEDA
PISCES
TRIANGULUM
Algol
Capella   Alpha Aurigids   PERSEUS

London – Looking North

August 01	21:00 (22:00 DST)
August 15	20:00 (21:00 DST)
September 01	19:00 (20:00 DST)
September 15	18:00 (19:00 DST)

-1   0   1   2   3   4   5

PISCES

M31
Andromeda Galaxy

DRACO

ANDROMEDA

CEPHEUS

LACERTA

Deneb

CYGNUS

Vega

LYRA

PEGASUS

Summer Triangle

M57
Ring Nebula

VULPECULA

M27
Dumbbell Nebula

SAGITTA

PISCES

M15

DELPHINUS

Altair

SERPENS CAUDA

EQUULEUS

E

AQUARIUS

AQUILA

M11
Wild Duck Cluster

SCUTUM

CAPRICORNUS

M22

M8
Lagoon Nebula

SAGITTARIUS

SUMMER SKY

June 01	01:00 (02:00 DST)
June 15	24:00 (01:00 DST)
July 01	23:00 (24:00 DST)
July 15	22:00 (23:00 DST)

CORONA AUSTRALIS

M7

Stellar magnitudes

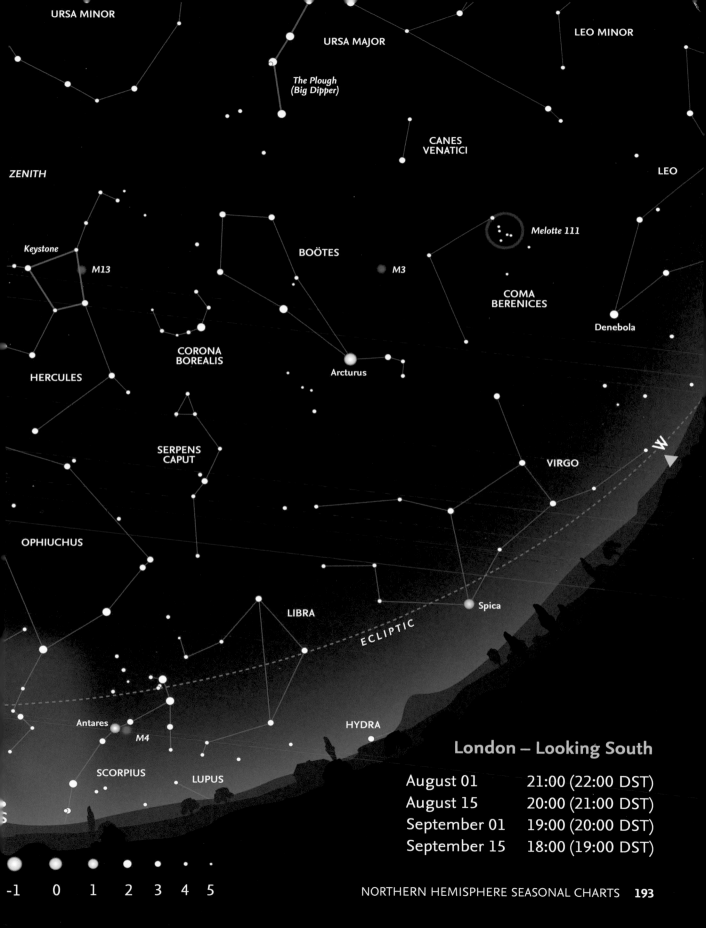

URSA MINOR

URSA MAJOR

LEO MINOR

The Plough
(Big Dipper)

CANES
VENATICI

ZENITH

LEO

Melotte 111

Keystone

M13

BOÖTES

M3

COMA
BERENICES

Denebola

CORONA
BOREALIS

Arcturus

HERCULES

SERPENS
CAPUT

VIRGO

W

OPHIUCHUS

Spica

LIBRA

ECLIPTIC

HYDRA

Antares

M4

London – Looking South

SCORPIUS

LUPUS

August 01	21:00 (22:00 DST)
August 15	20:00 (21:00 DST)
September 01	19:00 (20:00 DST)
September 15	18:00 (19:00 DST)

-1   0   1   2   3   4   5

EQUULEUS

PEGASUS

DELPHINUS

AQUILA

Altair

SAGITTA

LACERTA

ZENITH

M27
Dumbbell Nebula

Deneb

M11
Wild Duck Cluster

VULPECULA

CYGNUS

SCUTUM

SERPENS
CAUDA

M57
Ring Nebula

Summer
Triangle

CEPHEUS

LYRA

Vega

OPHIUCHUS

DRACO

URSA
MINOR

Keystone

HERCULES

M13

CORONA
BOREALIS

The Plough
(Big Dipper)

AUTUMN SKY

September 01	01:00 (02:00 DST)
September 15	24:00 (01:00 DST)
October 01	23:00 (24:00 DST)
October 15	22:00 (23:00 DST)

BOÖTES

CANES VENATICI

Stellar magnitudes:

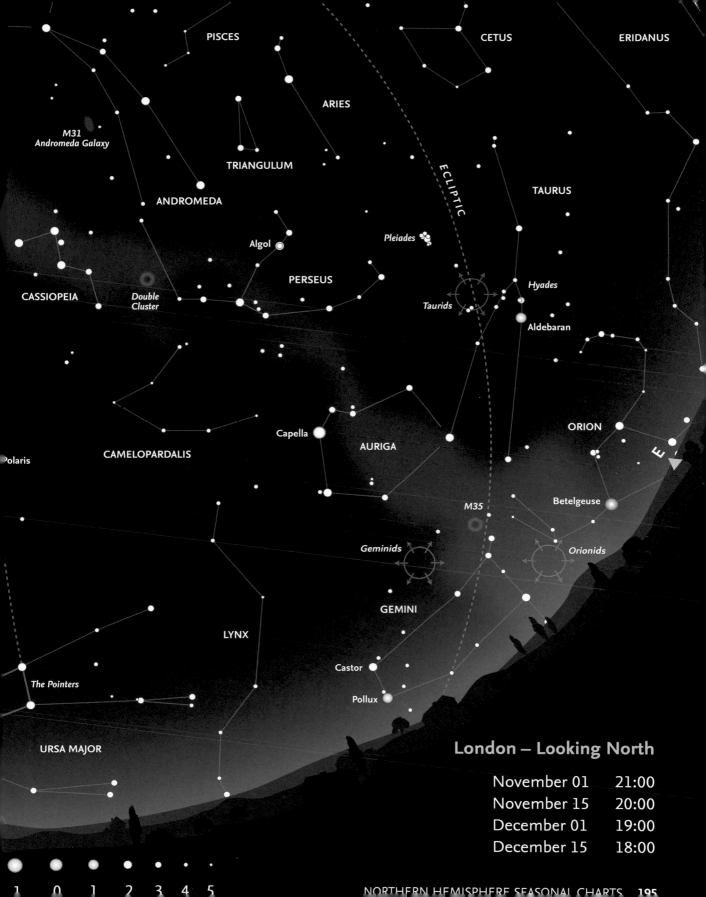

PISCES

CETUS

ERIDANUS

ARIES

M31
Andromeda Galaxy

TRIANGULUM

TAURUS

ANDROMEDA

ECLIPTIC

Algol

Pleiades

PERSEUS

Hyades

CASSIOPEIA

Double
Cluster

Taurids

Aldebaran

ORION

Capella

AURIGA

Betelgeuse

CAMELOPARDALIS

M35

Polaris

Orionids

Geminids

GEMINI

LYNX

Castor

Pollux

URSA MAJOR

The Pointers

London – Looking North

November 01	21:00
November 15	20:00
December 01	19:00
December 15	18:00

-1   0   1   2   3   4   5

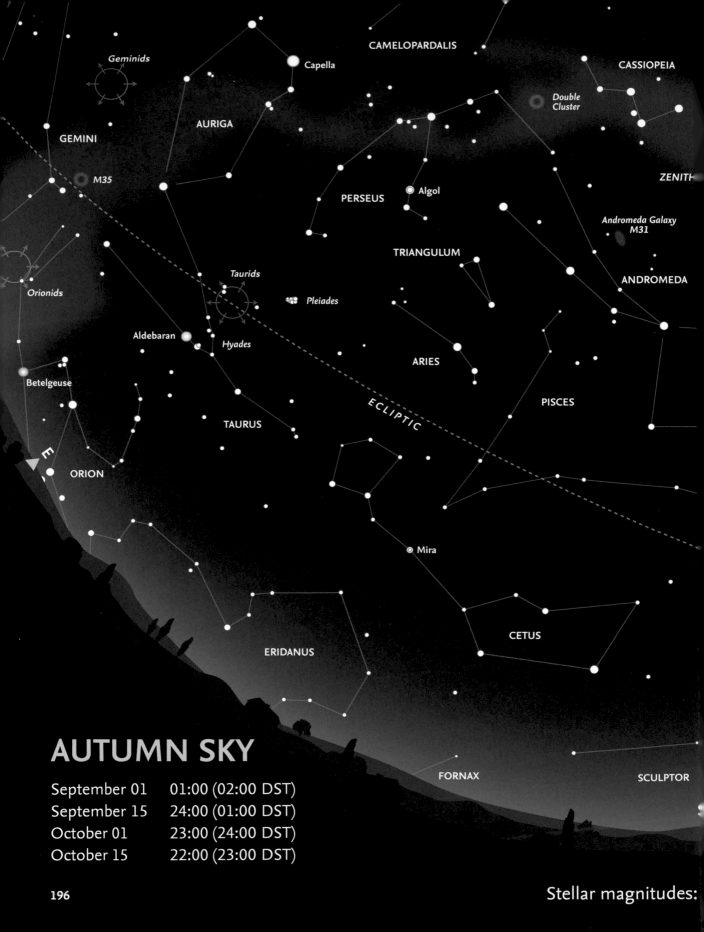

CAMELOPARDALIS

Geminids

Capella

CASSIOPEIA

Double
Cluster

AURIGA

GEMINI

ZENITH

M35

PERSEUS

Algol

Andromeda Galaxy
M31

TRIANGULUM

ANDROMEDA

Taurids

Orionids

Pleiades

Aldebaran

Hyades

ARIES

PISCES

Betelgeuse

ECLIPTIC

ORION

TAURUS

Mira

CETUS

ERIDANUS

AUTUMN SKY

September 01	01:00 (02:00 DST)
September 15	24:00 (01:00 DST)
October 01	23:00 (24:00 DST)
October 15	22:00 (23:00 DST)

FORNAX

SCULPTOR

Stellar magnitudes:

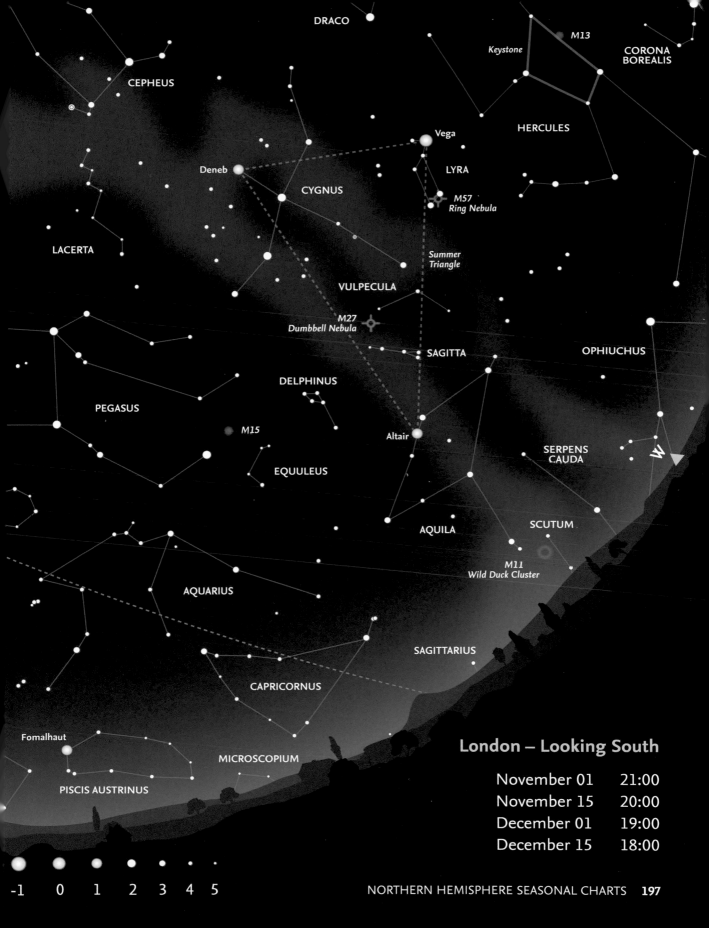

DRACO

CEPHEUS

Keystone   M13

CORONA
BOREALIS

Vega

HERCULES

Deneb

LYRA

CYGNUS

M57
Ring Nebula

LACERTA

*Summer
Triangle*

VULPECULA

OPHIUCHUS

M27
Dumbbell Nebula

SAGITTA

DELPHINUS

PEGASUS

M15

EQUULEUS

Altair

SERPENS
CAUDA

W

AQUILA

SCUTUM

M11
Wild Duck Cluster

AQUARIUS

SAGITTARIUS

CAPRICORNUS

Fomalhaut

MICROSCOPIUM

London – Looking South

November 01	21:00
November 15	20:00
December 01	19:00
December 15	18:00

PISCIS AUSTRINUS

-1   0   1   2   3   4   5

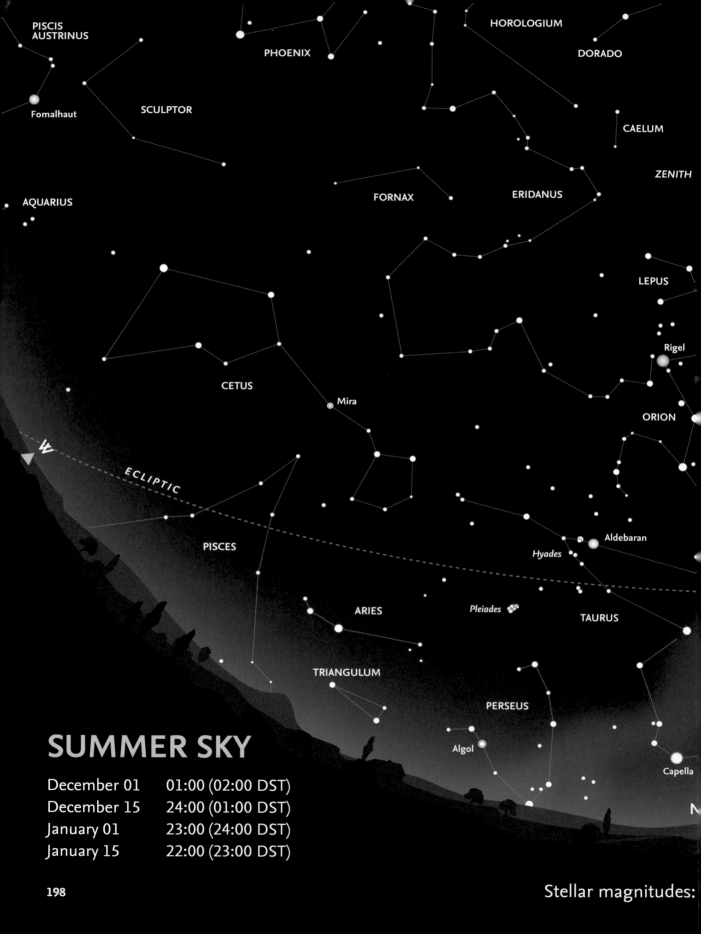

PISCIS
AUSTRINUS

Fomalhaut

SCULPTOR

AQUARIUS

PHOENIX

HOROLOGIUM

DORADO

CAELUM

ZENITH

FORNAX

ERIDANUS

CETUS

LEPUS

Rigel

Mira

ORION

ECLIPTIC

PISCES

ARIES

Hyades

Aldebaran

TAURUS

Pleiades

TRIANGULUM

PERSEUS

Algol

Capella

SUMMER SKY

December 01	01:00 (02:00 DST)
December 15	24:00 (01:00 DST)
January 01	23:00 (24:00 DST)
January 15	22:00 (23:00 DST)

Stellar magnitudes:

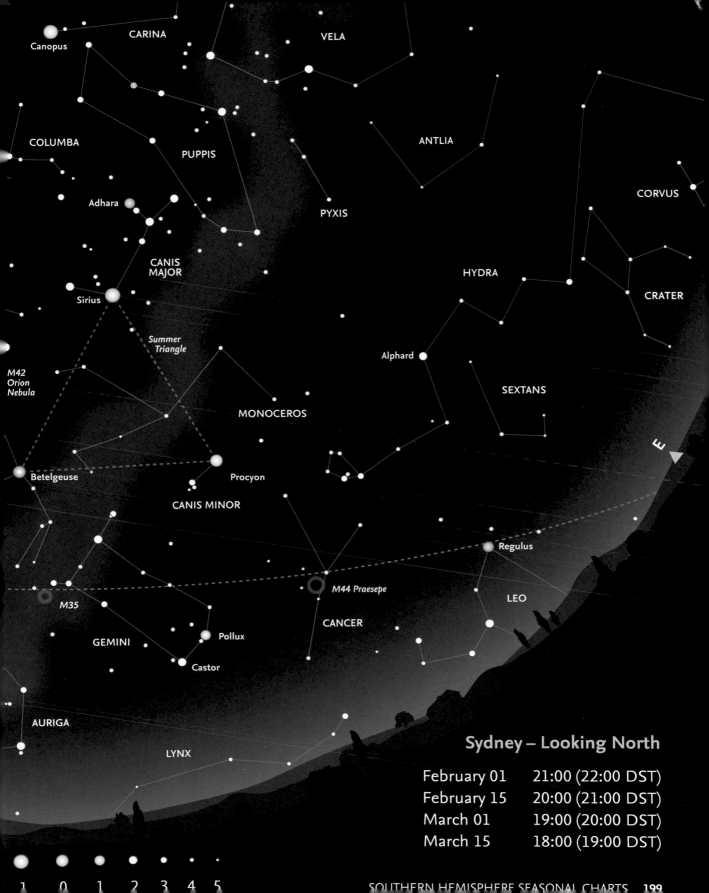

CARINA

Canopus

VELA

COLUMBA

PUPPIS

ANTLIA

PYXIS

CORVUS

Adhara

CANIS
MAJOR

HYDRA

CRATER

Sirius

*Summer
Triangle*

Alphard

SEXTANS

M42
Orion
Nebula

MONOCEROS

Betelgeuse

Procyon

CANIS MINOR

Regulus

M44 *Praesepe*

LEO

M35

CANCER

GEMINI

Pollux

Castor

AURIGA

LYNX

Sydney – Looking North

February 01	21:00 (22:00 DST)
February 15	20:00 (21:00 DST)
March 01	19:00 (20:00 DST)
March 15	18:00 (19:00 DST)

1 0 1 2 3 4 5

CANCER

ECLIPTIC

LEO

Regulus

Alphard

SEXTANS

HYDRA

ANTLIA

CRATER

E

CORVUS

MONOCEROS

Sirius

CANIS MAJOR

Adhara

COLUMBA

PUPPIS

PYXIS

VELA

CARINA

PICTOR

Canopus

VOLANS

Tarantula
Nebula

Eta Carinae Nebula

Southern
Pleiades

CHAMAELEON

CRUX

MUSCA

Acrux

CENTAURUS

Mimosa

Omega
Centauri

Hadar

Rigil Kent

CIRCINUS

LUPUS

ARA

APUS

TRIANGULUM
AUSTRALE

SUMMER SKY

December 01	01:00 (02:00 DST)
December 15	24:00 (01:00 DST)
January 01	23:00 (24:00 DST)
January 15	22:00 (23:00 DST)

Stellar magnitudes:

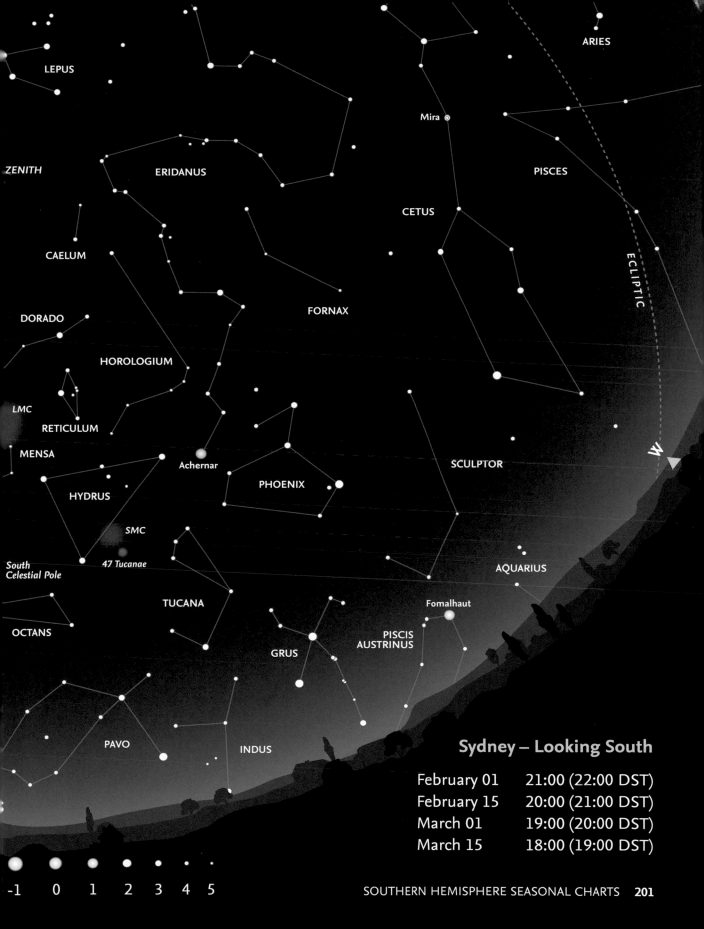

LEPUS

ZENITH

ERIDANUS

CAELUM

DORADO

HOROLOGIUM

LMC

RETICULUM

MENSA

Achernar

HYDRUS

SMC

South
Celestial Pole

47 Tucanae

TUCANA

OCTANS

GRUS

PAVO

INDUS

ARIES

Mira ◉

PISCES

CETUS

FORNAX

SCULPTOR

PHOENIX

AQUARIUS

Fomalhaut

PISCIS
AUSTRINUS

ECLIPTIC

Sydney – Looking South

February 01	21:00 (22:00 DST)
February 15	20:00 (21:00 DST)
March 01	19:00 (20:00 DST)
March 15	18:00 (19:00 DST)

-1   0   1   2   3   4   5

ERIDANUS

COLUMBA

VELA

Adhara

CANIS MAJOR

PUPPIS

ANTLIA

ZENITH

PYXIS

LEPUS

CRATER

Sirius

ORION

Alphard

MONOCEROS

SEXTANS

M42
Orion Nebula

HYDRA

W

ECLIPTIC

Procyon

Regulus

CANIS MINOR

LEO

M44
Praesepe

GEMINI

CANCER

LEO MINOR

Pollux

Castor

LYNX

AUTUMN SKY

March 01	01:00 (02:00 DST)
March 15	24:00 (01:00 DST)
April 01	23:00 (24:00 DST)
April 15	22:00

URSA MAJOR

N

Stellar magnitudes:

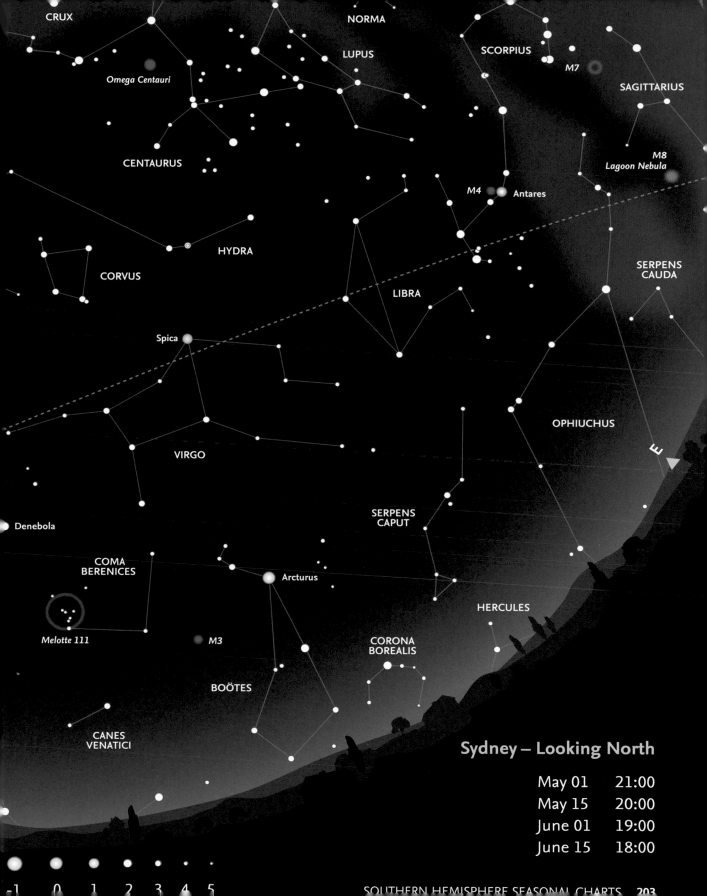

CRUX

NORMA

SCORPIUS

SAGITTARIUS

M7

Omega Centauri

LUPUS

M8
Lagoon Nebula

M4   Antares

CENTAURUS

HYDRA

SERPENS
CAUDA

CORVUS

LIBRA

Spica

OPHIUCHUS

VIRGO

E

Denebola

SERPENS
CAPUT

COMA
BERENICES

Arcturus

Melotte 111

M3

HERCULES

BOÖTES

CORONA
BOREALIS

CANES
VENATICI

Sydney – Looking North

May 01	21:00
May 15	20:00
June 01	19:00
June 15	18:00

-1   0   1   2   3   4   5

AUTUMN SKY

March 01	01:00 (02:00 DST)
March 15	24:00 (01:00 DST)
April 01	23:00 (24:00 DST)
April 15	22:00

Stellar magnitudes:

CRATER

SEXTANS

HYDRA

Alphard

CANCER

GEMINI

Procyon

CANIS MINOR

ZENITH

ANTLIA

PYXIS

MONOCEROS

VELA

PUPPIS

CANIS MAJOR

Eta Carinae Nebula

Sirius

Adhara

outhern Pleiades

CARINA

ORION

M42 Orion Nebula

VOLANS

Canopus

COLUMBA

CHAMAELEON

LEPUS

Tarantula Nebula

South Celestial Pole

PICTOR

MENSA

LMC

DORADO

CAELUM

ERIDANUS

HYDRUS

RETICULUM

MC

47 Tucanae

HOROLOGIUM

PHOENIX

Achernar

Sydney – Looking South

May 01	21:00
May 15	20:00
June 01	19:00
June 15	18:00

-1   0   1   2   3   4   5

ANTLIA

Omega Centauri

CENTAURUS

NORMA

ARA

HYDRA

LUPUS

ZENITH

SCORPIUS

M4   Antares

CORVUS

LIBRA

CRATER

ECLIPTIC

OPHIUCHUS

Spica

VIRGO

SERPENS
CAPUT

HERCULES

Arcturus

Keystone

COMA
BERENICES

CORONA
BOREALIS

M13

M3

BOÖTES

DRACO

WINTER SKY

June 01	01:00
June 15	24:00
July 01	23:00
July 15	22:00

Stellar magnitudes:

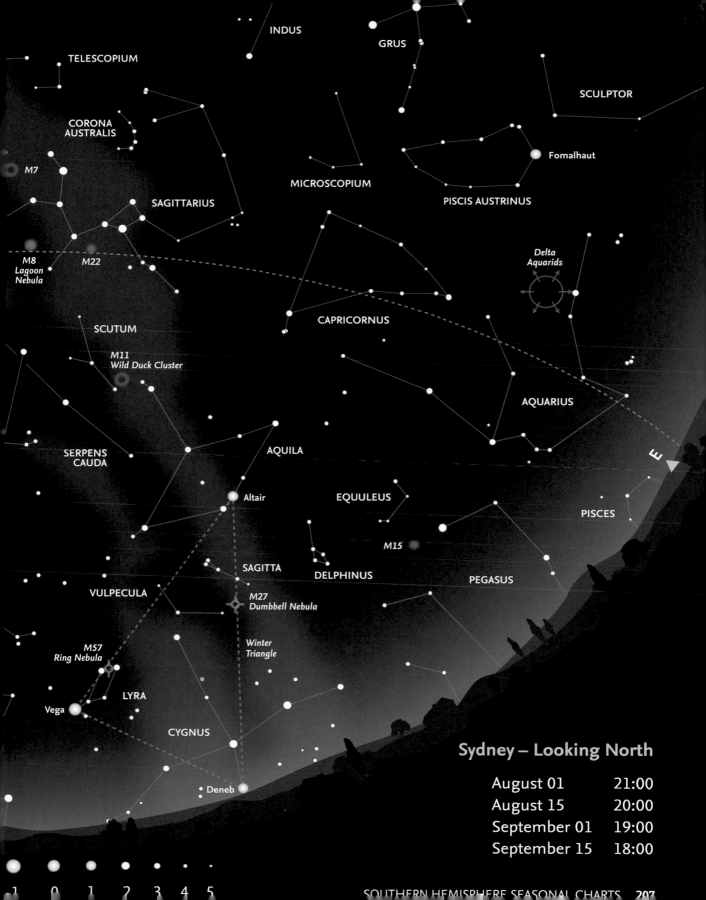

TELESCOPIUM

INDUS

GRUS

SCULPTOR

CORONA
AUSTRALIS

Fomalhaut

M7

MICROSCOPIUM

SAGITTARIUS

PISCIS AUSTRINUS

M8
Lagoon
Nebula

M22

Delta
Aquarids

CAPRICORNUS

SCUTUM

M11
Wild Duck Cluster

AQUARIUS

SERPENS
CAUDA

AQUILA

E

Altair

EQUULEUS

PISCES

M15

SAGITTA

DELPHINUS

PEGASUS

VULPECULA

M27
Dumbbell Nebula

M57
Ring Nebula

Winter
Triangle

LYRA

Vega

CYGNUS

Deneb

Sydney – Looking North

August 01	21:00
August 15	20:00
September 01	19:00
September 15	18:00

-1   0   1   2   3   4   5

AQUILA

M11 Wild Duck
Cluster

SCUTUM

SERPENS
CAUDA

EQUULEUS

M15

M22

M8 Lagoon
Nebula

PEGASUS

SAGITTARIUS

M7

AQUARIUS

CORONA
AUSTRALIS

CAPRICORNUS

MICROSCOPIUM

TELESCOPIUM

PISCES

*Delta
Aquarids*

INDUS

PAVO

PISCIS
AUSTRINUS

Fomalhaut

GRUS

OCTAN

TUCANA

South
Celestial Pole

SCULPTOR

47 Tucanae

SMC

HYDRUS

PHOENIX

Achernar

RETICULUM

ERIDANUS

LMC

HOROLOGIUM

DORADO

WINTER SKY

June 01	01:00
June 15	24:00
July 01	23:00
July 15	22:00

Stellar magnitudes:

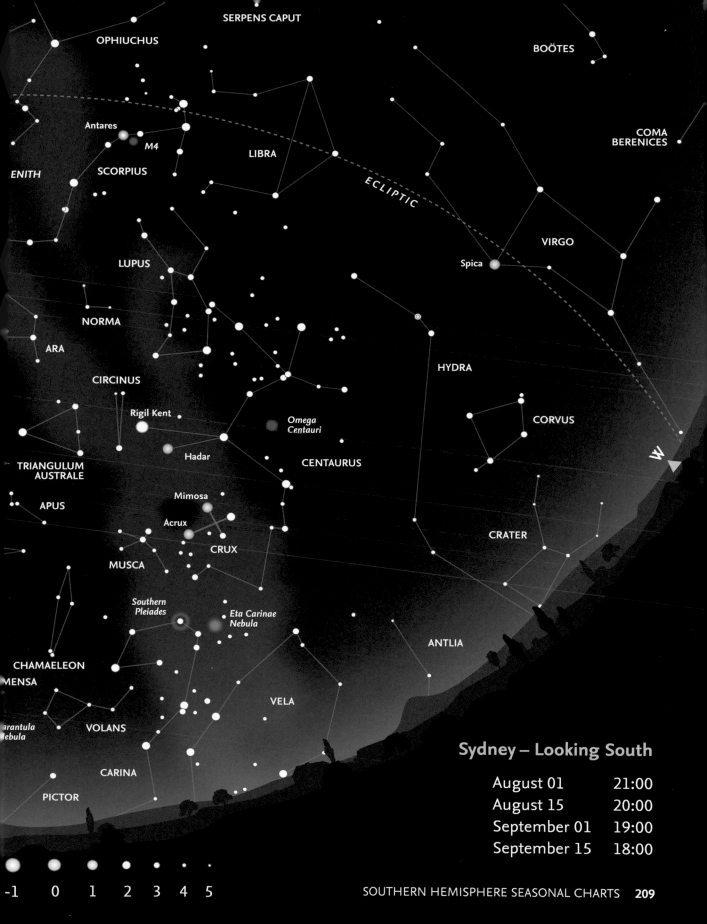

SERPENS CAPUT

OPHIUCHUS

BOÖTES

COMA
BERENICES

Antares

M4

LIBRA

ENITH

SCORPIUS

ECLIPTIC

VIRGO

Spica

LUPUS

NORMA

HYDRA

ARA

CIRCINUS

Rigil Kent

CORVUS

*Omega
Centauri*

TRIANGULUM
AUSTRALE

Hadar

CENTAURUS

CRATER

APUS

Mimosa

Acrux

CRUX

MUSCA

ANTLIA

*Southern
Pleiades*

*Eta Carinae
Nebula*

CHAMAELEON

MENSA

VELA

arantula
ebula

VOLANS

Sydney – Looking South

CARINA

August 01	21:00
August 15	20:00
September 01	19:00
September 15	18:00

PICTOR

-1   0   1   2   3   4   5

TELESCOPIUM

SCORPIUS

M7

INDUS

CORONA
AUSTRALIS

MICROSCOPIUM

GRUS

M8 Lagoon
Nebula

SAGITTARIUS

M22

OPHIUCHUS

PISCIS AUSTRINUS

Fomalhaut

CAPRICORNUS

SCUTUM

AQUARIUS

M11 Wild Duck
Cluster

EQUULEUS

SERPENS
CAUDA

M15

Altair

PEGASUS

AQUILA

DELPHINUS

SAGITTA

M27
Dumbbell Nebula

VULPECULA

CYGNUS

LACERTA

SPRING SKY

August 01	03:00
August 15	02:00
September 01	01:00
September 15	24:00

Deneb

Stellar magnitudes:

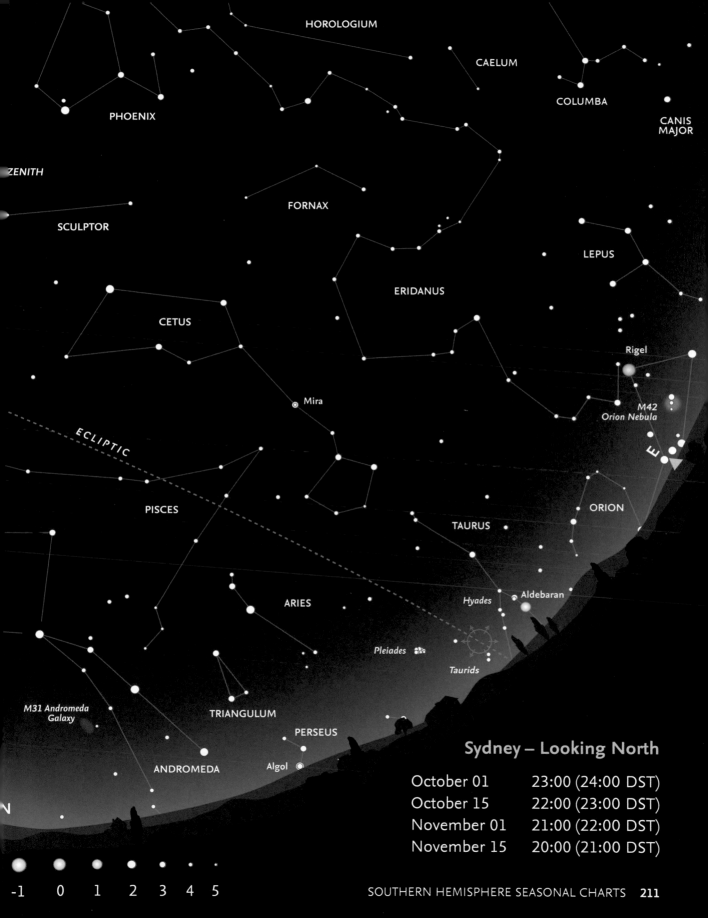

HOROLOGIUM

CAELUM

COLUMBA

CANIS
MAJOR

PHOENIX

ZENITH

SCULPTOR

FORNAX

LEPUS

ERIDANUS

Rigel

CETUS

Mira

M42
Orion Nebula

ECLIPTIC

PISCES

ORION

TAURUS

ARIES

Hyades   Aldebaran

Pleiades

Taurids

M31 Andromeda
Galaxy

TRIANGULUM

PERSEUS

ANDROMEDA   Algol

N

Sydney – Looking North

October 01	23:00 (24:00 DST)
October 15	22:00 (23:00 DST)
November 01	21:00 (22:00 DST)
November 15	20:00 (21:00 DST)

-1   0   1   2   3   4   5

Pleiades

TAURUS

Hyades

Mira

CETUS

SCULPTOR

ZENITH

PHOENIX

FORNAX

ERIDANUS

Achernar

ORION

Rigel

HOROLOGIUM

47 Tucanae

M42 Orion
Nebula

SMC

LEPUS

CAELUM

HYDRUS

DORADO

RETICULUM

COLUMBA

PICTOR

LMC

South
Celestial Pole

Tarantula
Nebula

MENSA

CANIS
MAJOR

Canopus

Adhara

VOLANS

CHAMAELEON

PUPPIS

CARINA

Southern
Pleiades

SPRING SKY

August 01	03:00
August 15	02:00
September 01	01:00
September 15	24:00

VELA

Eta Carinae
Nebula

Stellar magnitudes:

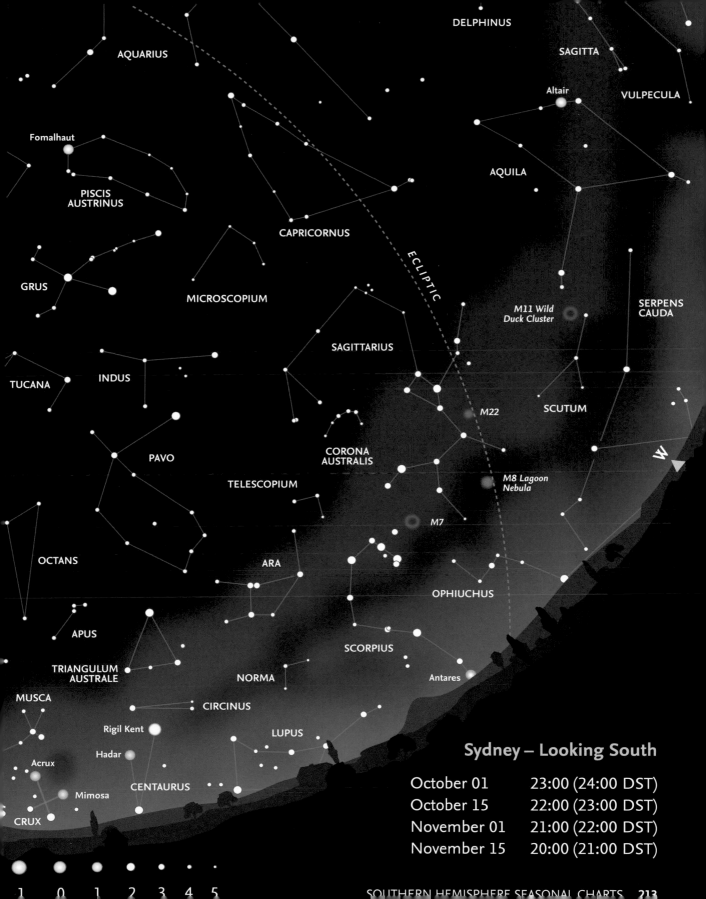

DELPHINUS

SAGITTA

VULPECULA

AQUARIUS

Altair

AQUILA

Fomalhaut

SERPENS
CAUDA

PISCIS
AUSTRINUS

M11 Wild
Duck Cluster

CAPRICORNUS

ECLIPTIC

SCUTUM

GRUS

MICROSCOPIUM

SAGITTARIUS

M22

TUCANA

INDUS

CORONA
AUSTRALIS

M8 Lagoon
Nebula

PAVO

M7

TELESCOPIUM

W

OCTANS

ARA

OPHIUCHUS

APUS

SCORPIUS

TRIANGULUM
AUSTRALE

NORMA

Antares

MUSCA

CIRCINUS

Rigil Kent

LUPUS

Acrux

Hadar

CENTAURUS

Mimosa

CRUX

Sydney – Looking South

October 01	23:00 (24:00 DST)
October 15	22:00 (23:00 DST)
November 01	21:00 (22:00 DST)
November 15	20:00 (21:00 DST)

1 0 1 2 3 4 5

FURTHER RESOURCES AND READING

Useful Starcharts and Apps to find your way around the Sky

Collins Planisphere
(2013) HarperCollins, London

Astronomical software for your computer:

Stellarium – www.stellarium.org

Celestia – www.shatters.net/celestia

Apps for your smartphone or tablet:

StarMap 3D+ has HD retina graphics for iOS devices

Mobile Observatory gives you a 3D view of the Solar System and has a calendar of celestial events

Spot the Station is a free mobile service that notifies you when the ISS is overhead from your location

Star Rover allows you to simulate eclipses, shows a realistic sunrise and sunset and lots of beautiful images

StarTracker finds your location via GPS and is simple to use

Sky Live is designed to help you plan your next stargazing session by supplying hourly weather conditions

Night Sky Pro also provides reports of weather conditions, and has a feature that shows you what the sky looks like in different types of light

Check the Weather

UK – www.metoffice.org.uk

US – www.weather.gov

Australia – www.bom.gov.au

The Sun and Moon

Current conditions of the Sun – http://spaceweather.com

Sunrise and moonrise times – www.timeanddate.com

For the British Isles, detailed information about the lunar calendar is available from the UK Hydrographical Office, a department of Her Majesty's Nautical Almanac Office – http://astro.ukho.gov.uk

More general information can be retrieved from the US Naval Observatory's Astronomical Applications Department: http://aa.usno.navy.mil

Upcoming Celestial Events

Dunlop, Storm & Tirion, Wil (published yearly) *Collins Guide to the Night Sky* HarperCollins, London

Comets and meteor showers:

To stay up to date with bright comets, visit https://in-the-sky.org/data/comets.php

For upcoming meteor showers visit http://imo.net – you can submit your own observations here

Lunar and solar eclipses:

Eclipse information was historically published by the Royal Observatory Greenwich in their annual edition of the Nautical Almanac. Today, this tradition continues, with information now available from Her Majesty's Nautical Almanac Office, a department of the UK Hydrographical Office, and the US Naval Observatory. To find an upcoming eclipse, visit: http://astro.ukho.gov.uk/eclipse/ or go to http://eclipse.gsfc.nasa.gov/eclipse.html

Planetary transits:

Dates for the next set of transits can be found here – http://eclipse.gsfc.nasa.gov/transit/transit.html

Information about Satellites passing overhead

Astronomical and space calendar – http://calsky.com

The International Space Station – https://spotthestation.nasa.gov

Dark Skies

You can find observational advice and ways to join the fight to reduce light pollution here – www.windows2universe.org/citizen_science/starcount

Find a dark sky reserve or park here – http://darksky.org/idsp/reserves

For the UK you can find Dark Sky Discovery sites at: www.darkskydiscovery.org.uk

GLOSSARY

altitude	The apparent height of an object above an observer's horizon, measured in degrees with the horizon at 0° and the zenith at 90°.
angular size	A measure of how wide an object in the sky appears, measured in degrees, arcminutes and arcseconds.
aphelion	The point in Earth's orbit when it is farthest from the Sun.
apogee	The point in a satellite's orbit (usually the Moon) when it is farthest from Earth.
arcminutes; arcseconds	A unit of angular distance. 60 arcminutes make up one degree; 60 arcseconds make up one arcminute.
asterism	A pattern of bright stars usually smaller than a constellation, but sometimes spanning multiple constellations. The Plough or Big Dipper is a famous northern hemisphere asterism.
astronomical unit	Defined as the average distance between the centres of the Sun and Earth, about 149,600,000 km (93,000,000 miles).
averted vision	The practice of viewing an object indirectly, making it appear brighter by avoiding the eye's blind spot when dark adapted.
axial tilt	The separation in degrees between the axis of a planet and an imaginary line perpendicular to its orbital plane.
axis	An imaginary line connecting the north and south poles through the centre of an object, such as the Earth or another planet.
azimuth	Measured in degrees — the bearing of an object local to the observer, where north is 0°, east is 90°, south is 180° and west is 270°.
cardinal directions	The directions north (N), east (E), south (S) and west (W).
celestial pole	The most northerly and southerly points on the celestial sphere.
celestial sphere	An imaginary sphere with Earth at the centre. The positions of the stars and other celestial objects are given as coordinates on this sphere.
circumpolar	Stars that revolve around the celestial pole and never set below the horizon as seen by the observer. The number of circumpolar stars depends on latitude.
constellation	A recognisable or traditional pattern of stars. There are 88 formally acknowledged constellations covering the whole sky. Each covers an area defined by a fixed boundary.
dark adapted	When the eye operates on mesopic or scotopic vision, with a dilated pupil. Occurs naturally in dimly lit conditions. Becoming fully dark adapted in near total darkness takes about 45 minutes.

declination	Equivalent to latitude on Earth – how far north or south an object is on the celestial sphere. Measured in degrees with the celestial equator at 0°.
dwarf planet	A spherical object that doesn't clear its orbit around the Sun.
eclipse	When the Moon moves in front of the Sun (solar eclipse) or moves into the shadow of Earth (lunar eclipse).
ecliptic	The Earth's orbital plane around the Sun projected into space (onto the celestial sphere).
elliptical	The orbits of the planets are elliptical rather than circular – their distance varies from the Sun.
equatorial plane	The Earth's equator extended out into space as a flat plane (onto the celestial sphere).
equinox (vernal, autumnal)	When the Earth is neither tilted towards nor away from the Sun. These occur in March and September, when people on the Earth experience an equal number of hours of day and night.
exposure time	Measured in seconds – the length of time a camera's shutter is left open. Longer exposure times are used to collect more light for a brighter photograph.
hour angle	Measured as an angle or period of time – the separation between an object's apparent position and the observer's local meridian line.
latitude	A geographical coordinate given in degrees: how far north or south somewhere is from the equator (which is 0°).
light pollution	Artificial light from towns and cities scattered and reflected by the atmosphere, resulting in a brightened, orange sky through which fainter stars and other objects are more difficult to see.
light-year	A unit of distance in astronomy – the distance light travels in one year, about 9,500,000,000,000 km (6,000,000,000,000 miles).
limiting visual magnitude	Dictates the magnitude of the faintest star visible to the unaided eye
local meridian	An imaginary line across the sky, connecting the north and south cardinal points, and crossing the zenith.
longitude	A geographical coordinate given in degrees: how far east or west somewhere is from the Prime Meridian (which is 0°).
lunar crater	A recess in the lunar surface created by the impact of a meteorite, such as an asteroid.
lunar sea	Latin: mare. Dark region on the Moon made of basalt-rich solidified lava from ancient volcanic activity.
magnitude	The apparent brightness of an object in the sky, where smaller (and negative) values indicate greater brightness.

mesopic vision	A state in which the eye is partially dark adapted, using a combination of rod and cone cells, allowing fainter objects to be seen by the observer. The most common form of human night vision.
occultation	When an object is hidden behind another object in the sky. For example, Jupiter can occult its moons as they pass behind the planet as seen from Earth.
orbital plane	An imaginary plane defined by the orbital path of any object in space.
penumbra	A lighter region around the umbra of a planet or moon's shadow.
perigee	The point in a satellite's orbit (usually the Moon) when it is closest to Earth.
perihelion	The point in Earth's orbit when it is closest to the Sun.
photopic vision	The ordinary state of the eye in well-lit conditions. Cone cells are active, allowing colours to be perceived.
planisphere	A simple tool made from two discs that indicates the appearance of the sky from a given latitude for any night of the year.
precession	The slow wobble of Earth's axis resulting in the gradual westward drift of the First Point of Aries.
prime meridian	An imaginary line on Earth's surface connecting the north and south poles, and passing through the Royal Observatory Greenwich. It marks 0° of longitude.
projection	A flat, warped representation of a curved surface. For example, an all-sky map represents the celestial sphere stretched onto a rectangle.
Rayleigh scattering	Named after British physicist Lord Rayleigh. The phenomenon in which blue light is preferentially scattered by oxygen and nitrogen in the atmosphere, resulting in a blue sky.
refraction	The change in direction of light waves as they propagate through a medium, or change medium, such as when entering Earth's atmosphere from the vacuum of space.
resolving power	The ability for a telescope to separate and distinguish features and objects. High resolving power is required to see features on planets, or separate close pairs of stars.
retrograde motion	The apparent change of motion of a planet in the opposite direction to that of other bodies orbiting the Sun, as seen from the perspective of Earth.
right ascension	A coordinate on the celestial sphere describing how far east or west of the First Point of Aries an object is. Conventional units are hours, minutes and seconds, increasing in the direction of east.

scotopic vision	A state in which the eye is optimally dark adapted, operating almost entirely on rod cells, allowing much fainter objects to be seen by the observer.
season	In astronomy, a season is one quarter of Earth's orbit. Each season starts at its corresponding solstice or equinox. Opposite hemispheres experience opposite seasons.
seeing	A measure of the effect of atmospheric turbulence on the steadiness of the appearance of the stars and other objects. Steady conditions are described as good seeing.
sidereal day	The period of Earth's rotation measured against the fixed stars. A mean sidereal day is 23 hours, 56 minutes and 4.09 seconds long.
solar day	The period of Earth's rotation measured against the Sun. A mean solar day is 24 hours long.
solstice	A point in Earth's orbit at which its axis points directly to and from the Sun. Solstices occur in June and December, marking the beginnings of astronomical summer or winter depending on the hemisphere.
supermoon	Slightly larger apparent size of the Moon when it is at perigee.
synodic lunar period	The time between two successive Full Moons, equal to 29.5 days.
transit	The passing of one astronomical object in front of a larger one. For example, Mercury and Venus can transit the Sun as seen from Earth. Jupiter's moons can transit the planet, casting shadows on its clouds.
transparency	A measure of the clarity of the atmosphere and its effect on the appearance of stars and other objects. Good transparency allows fainter objects to be seen more clearly, as well as more vivid colour.
umbra	The darkest, central region of a planet or moon's shadow in space.
zenith	The point immediately above the observer, with an altitude of 90°.
zodiac	A group of 13 constellations through which the Sun traces an apparent path each year called the ecliptic.

OBSERVING LOG

Date:.................................... Seeing:............................... Transparency:.....................

Object and coordinates used to set RA circle:.............................

Objects:.................................

Coordinates:..........................

Eyepiece: ....

Magnification:......................

Notes:....................................

....

....

....

....

Date:............................ Seeing:............................ Transparency:............................

Object and coordinates used to set RA circle:............................

Objects:............................

Coordinates:............................

Eyepiece:............................

Magnification:............................

Notes:............................

....

....

....

....

ACKNOWLEDGMENTS

Star Charts by Wil Tirion – www.wil-tirion.com.
Constellation charts 116–171, Seasonal charts 182–219. Illustrations pages 15, 17 bottom, 21 bottom, 22–23, 25 bottom, 26–27, 28–29, 30 top, 31, 35, 39 top, 55.

Illustrations

38, shooarts/Shutterstock; 51 top, Panacea Doll/Shutterstock; 51 bottom, Designua/Shutterstock; 52 top, Tom Kerrs; 54, © Tom Kerss; 86, mike mols/Shutterstock; 96 top, Sergey_R/Shutterstock; 112, NASA/JPL-Caltech/R. Hurt (SSC-Caltech).
All other illustrations created by HarperCollins Publishers.
The following elements were used: 10, 12, 14, 15 bottom, 19, 20, 34, 36–37, 42 top, 43, Sun: xfox01/Shutterstock; 10, 17, 20, 21 top, 25 top, 34, 42, 46, Earth: Antony McAuley/Shutterstock; 12, 36–37, 46, Planets: Antony McAuley/Shutterstock; 14, 19, Earth: NASA; 34, 42, 43, Moon illustration: Koryaba/Shutterstock; 34, Moon photograph: Lick Observatory; 36–37, Haumea: NASA; Ceres: NASA/JPL/CIT; 36–37 other elements, shooarts/Shutterstock; 43, hand: Hein Nouwens/Shutterstock; face: Skitale/Shutterstock; coin: AlexanderZam/Shutterstock; 44 bottom, Fouad A. Saad/Shutterstock; 46 top, rainbow: veronchick84/Shutterstock; 45, 46 top, Earth: CarpathianPrince/Shutterstock; 53 top, Tom Kerss; 87 top, Trikona/Shutterstock.

Photographs

Introduction 6–7: McCarthy's PhotoWorks/Shutterstock.
The Night Sky: 8–9, Nook Thitipat/Shutterstock; 11, © Tom Kerss; 16 top, Sean Pavone/Shutterstock; 16 bottom, BanjerdPhoto/Shutterstock; 31 bottom, © Tom Kerss; 33, ESO/Y. Beletsky; 38 bottom, © Tom Kerss; 40–41 Tuanna2010:CC by 3.0; 44 top, Pat_Hastings/Shutterstock; 44 middle, © Radmila Topalovic; 45 bottom, NASA/JPL/Texas A&M/Cornell, 47; © Tunç Tezel.
Planning your Stargazing Session: 48–49, Viktar Malyshchyts/Shutterstock; 52 bottom left, Chris Shur; 52 bottom right, © Tom Kerss; 53 bottom, © Tom Kerss; 58; National Maritime Museum, Greenwich, London, Caird Collection; 56–57, Collins Bartholomew Ltd; 58, Stellarium; 60, David Hughes; 61, © Radmila Topalovic; 62–63, © CBW / Alamy Stock Photo; 64, © Tom Kerss; 65, EpicStockMedia/Shutterstock.
Start with your Eyes: 66–67, eugenegurkov/Shutterstock; 68, Jeremy Stanley: CC by 2.0; 69, © Radmila Topalovic; 70, Philipp Salzgeber: CC by-SA 2.0 Austria; 72, Steve Jurvetson: CC by 2.0; 73, saraporn/Shutterstock; 74–75, G. Hüdepohl (atacamaphoto.com).
Taking Pictures: 76–77, Ververidis Vasilis/Shutterstock; 79 top left, Bairachnyi Dmitry/Shutterstock; 79 top right, © Tom Kerss; 79 bottom, Anthony Guiller E. Urbano (www.nightskyinfocus.com); 80 top left, © Tom Kersss; 80 top right, NASA,ESA, M. Robberto (Space Telescope Science Institute/ESA) and the Hubble Space Telescope Orion Treasury Project Team; 80 bottom, © Tom Kerss; 81, © Brendan Owens.
Using Binoculars or a Telescope: 82–83, PlanilAstro/Shutterstock; 84 bottom, bDimedrol68/Shutterstock; 84 top, Anton Mykhailovskyi/Shutterstock; 85 top, Jan Sandberg www.desert-astro.com; 85 bottom, © Tom Kerss; 86, photoHare/Shutterstock; 87 top, mike mols/shutterstock; 88, © Tom Kerss; 89 top left and bottom, © Tom Kerss; 89 top right, Ralf Vandebergh: CC by-SA 3.0.
Things to See: 90–91, Bill Frische/Shuttesrtock; 93, NASA/GSFC/Arizona State University; 94 top left, NASA/GSFC/Arizona State University; 94 bottom left, © Radmila Topalovic; 94 right, © Tom Kerss; 95 top left, © Damian Peach; 95 top right, © Tom Kerss; 95 bottom, © Tom Kerss; 96 bottom, © Tom Kerss; 97 top, © Tom Kerss; 97 bottom, Tim Valk; 99 left, © Tom Kerss; 99 right, © Damian Peach; 100–101, chrisdorney/Shutterstock.com; 102, NASA/JSC; 103, E. Kolmhofer, H. Raab; Johannes-Kepler-Observatory,

Linz, Austria (www.sternwarte.at): CC by-SA 3.0; 104, © Tom Kerss; 105, © Tom Kerss; 107 top, THANAKRIT SANTIKUNAPORN/Shutterstock; 107 bottom left, Zhorov Igor/Shutterstock; 107 bottom right, John Lord: CC by2.0; 108, © Tom Kerss; 109, © Tom Kerss; 110 bottom, © Tom Kerss; 111 bottom, © Tom Kerss; 111 top, NASA/SCIENCE PHOTO LIBRARY; 113, NASA/JPL-Caltech/R. Hurt (SSC-Caltech).

Constellations and Seasonal Objects: 114–115, Reinhold Wittich/Shutterstock; 118 left, isak55/Shutterstock; 118 right, J. Schedler (Panther Observatory); 119, NASA, ESA, HEIC, and The Hubble Heritage Team (STScI/AURA); 120 top, Reinhold Wittich/Shutterstock; 120 bottom, Henryk Kowalewski: CC by-SA 2.5; 121, Image Data – Subaru Telescope (NAOJ), Hubble Legacy Archive; Processing – Robert Gendler; 122, Digitized Sky Survey 2. Acknowledgment: Tom Kerss; 123 top, Tragoolchitr Jittasaiyapan/Shutterstock; 123 bottom, Matipon/Shuttrstock; 124 top, Digitized Sky Survey – STScI/NASA, Colored & Healpixed by CDS/ Aladin Sky Atlas; 124 bottom, ESO/INAF-VST/OmegaCAM. Acknowledgement: OmegaCen/Astro-WISE/Kapteyn Institute; 125, Tragoolchitr Jittasaiyapan/Shutterstock; 126–127, ESO/INAF-VST/OmegaCAM. Acknowledgement: OmegaCen/Astro-WISE/Kapteyn Institute; 128, Atlas Image courtesy of 2MASS/UMass/IPAC-Caltech/NASA/NSF; 129 left, McCarthy's PhotoWorks/Shutterstock; 129 right, Jschulman555: CC by-SA 4.0; 130, Atlas Image courtesy of 2MASS/UMass/IPAC-Caltech/NASA/NSF; 131 top, Antares_StarExplorer/Shutterstock; 131 bottom, © Tom Kerss; 132 top, Antares_StarExplorer/Shutterstock; 132 bottom, Digitized Sky Survey 2. Acknowledgment: Tom Kerss; 133, Adam Block/Mount Lemmon SkyCenter/University of Arizona: CC by-SA 3.0; 134 left, Reinhold Wittich/Shutterstock; 134 right, Todd Vance: CC by-SA 2.5; 135, Viktar Malyshchyts/Shutterstock; 136 left, peresanz/Shutterstock; 136 right, Roberto Mura: CC by-SA 3.0; 137, NASA, ESA; 138–139, Antares_StarExplorer/Shutterstock; 140 left, Allexxandar/Shutterstock; 140 right, B. Balick, J. Alexander (University of Washington), et al., NASA; 141, Albert Barr/Shutterstock; 142 left, The Hubble Heritage Team (AURA/STScI/NASA); 143 right, © Tom Kerss; 143, Igor Chekalin/Shutterstock; 144 left, Albert Barr/Shutterstock; 144 right, J. Jongmans/Shutterstock; 145, Tragoolchitr Jittasaiyapan/Shutterstock; 146 bottom, Digitized Sky Survey 2. Acknowledgment: Davide De Martin; 147, ESO/J. Emerson/VISTA. Acknowledgment: Cambridge Astronomical Survey Unit; 146 top, Adam Block/Mount Lemmon SkyCenter/University of Arizona: CC by-SA 3.0; 148, © Damien Lemay – damien.lemay@globetrotter.net; 149 top, © Tom Kerss; 149 bottom, Digitized Sky Survey - STScI/NASA, Colored & Healpixed by CDS/ Aladin Sky Atlas; 150 top, Tragoolchitr Jittasaiyapan/Shutterstock; 150 bottom, ESA/Hubble & NASA; 151, Allexxandar/Shutterstock; 152–153, ESA/Hubble & NASA; 154, ESA/Hubble & NASA: CC by 3.0; 155 left, This publication makes use of data products from the Two Micron All Sky Survey, which is a joint project of the University of Massachusetts and the Infrared Processing and Analysis Center/California Institute of Technology, funded by the National Aeronautics and Space Administration and the National Science Foundation.; 155 right, ESO; 156 left, Tragoolchitr Jittasaiyapan/Shutterstock; 156 right, ESO; 157, Tragoolchitr Jittasaiyapan/Shutterstock; 158 top, Atlas Image courtesy of 2MASS/UMass/IPAC-Caltech/NASA/NSF; 158 bottom, ESO; 159, Digitized Sky Survey - STScI/NASA, Colored & Healpixed by CDS/ Aladin Sky Atlas; 160, NASA, ESA, and the Hubble Heritage Team (STScI/AURA); 161 top, w:en:Palomar Observatory/w:en:STScI/w:en:WikiSky; 161 bottom, Roberto Mura: CC by-SA 3.0; 162 left, ESO/B. Tafreshi (twanight.org); 162 right, Roberto Mura; 163, ESO/G. Beccari; 164–165, Matipon/Shutterstock; 166 top, ESO; 166 bottom, Tom Kerss & Digitized Sky Survey 2; 167, ESO; 168 top, ESO; 168 bottom, ESO/Y. Beletsky; 169, Digitized Sky Survey - STScI/NASA, Colored & Healpixed by CDS/ Aladin Sky Atlas; 170 left, ESA/NASA/JPL-Caltech/STScI; 170 right, ESO/M.-R. Cioni/VISTA Magellanic Cloud survey. Acknowledgment: Cambridge Astronomical Survey Unit; 171, John A Davis/Shutterstock.

Start Stargazing!: 72–173, Nejron Photo/Shutterstock; 174–175, ESO/B. Tafreshi (twanight.org); 176–177, Josef Pittner/Shutterstock; 178–179, ESO/M.-R. Cioni/VISTA Magellanic Cloud survey. Acknowledgment: Cambridge Astronomical Survey Unit.

Seasonal Charts: 180–181, ESO/B. Tafreshi (twanight.org).